The Corporation

or

I Didn't Set Out To Be Successful

or

One Man's Experience of Navigating the World of Engineers (Not the Train-Driving Kind)

By Jack Martino

Immacolata

First Edition published in 2025

ISBN: 978-1-7635795-1-4

This book was written on aeroplanes. Some of it was also written in airports and houses but mostly on aeroplanes. Aeroplanes so old that they didn't have TVs in the back of the seats or a wireless entertainment system. Given the choice, I would watch a movie, even a bad movie, over writing. But my weekly commute to a worksite (two hours each way) gave me lots of time for pondering. And ponder I did. I found myself telling stories about how bad my bosses had been over the years. I think that I've worked for three genuinely good bosses in my career and the rest were either incompetent, self-serving or sociopathic (commonly all three). I have written about them for posterity; for the benefit of future generations, who will probably work and live in a utopian existence and never experience bad management first hand. I do not expect that this will become a text book.

The names of *the corporations* referred to herein are pseudonyms. This is because I don't want to be identified and sued.

This is a true story. There are only minor embellishments here and there when I can't actually remember the details or to make me look good. However, to protect the innocent, all the characters named on the following pages are pseudonyms.

I have enjoyed this story twice: the first time when I lived it out and the second when I wrote it down. I hope that someone enjoys reading it.

Jack Martino, BEng(Hons), CPEng, NPER

(Jack Martino is a pseudonym)

For my darling wife, Bella (thank you for being awesome and proof reading) and for my precious children, Michaela, Rafael and Emily.

(Bella, Michaela, Rafael and Emily are pseudonyms)

CONTENTS

PROLOGUE

I was at a neighbourhood Christmas party talking to Wranger (I'd met him once before out the front of our houses when a particularly large blue moon was advertised as a once in a lifetime experience). We live on a nice street, in a house with a pool and nice views, and presumably nice neighbours.

"So, Jack, what do you do?"

I find this seemingly innocuous question difficult to answer. I feel a bit guilty if I say *"I'm an engineer,"* because although I have an engineering degree, I haven't done a calculation for about two decades. I think a more accurate response would be: *"I'm a failed engineer... a manager!"* but this tends to go over the head of normal people (non-engineers). And a vague but truthful *"It's complicated,"* is a massive conversation killer. Wranger seemed like a pretty good bloke, with a full glass in his hand, so I went with the long answer:

"Well, Wranger, if I want to be a total twit, I say (*deep voice with an English accent*) 'I advise the boards of international mining companies regarding their capital investment decisions.' And if I don't want to be a total twit, I say 'I review projects.'"

How did I get here? Not to the neighbourhood Christmas Party – that was an easy walk two doors away. How did I get to travel around the world 'reviewing projects?' It seems like only yesterday that I was in my first job. I had a bit of an alter-ego in those days, sort of like Batman. By day, I wore normal clothes and went to university to learn engineering stuff. But by night, I donned crappier clothes and stacked shelves at the local supermarket. Night Fill could be tough – if you got the freezer aisle, you needed gloves. I remember the first time that I had to go to the toilet while on night fill. I was overcome with joy at the realisation that I was getting paid to dump. Not very much, since my hourly rate was about $12. As the years have gone on, dumps at work have progressively been better remunerated. But as an international consultant, it's all wrapped up in the day rate.

How did I get here? Read on, my friend...

PREPARATION (EDUCATION)

When middle class people finish school, they generally go to university in order to be considered for acceptance into *the corporation*. People who are either at the top or the bottom seem to fish out different opportunities in life that don't involve higher education: Buy a business, work for your dad's business or go to trade school (which also generally involves owning your own business soon after).

I was a pretty good student (this is how students with an A average describe themselves). Dad was an engineer and Mum was a languages teacher. This meant that I always had 'back up teachers' at home. Maths, science and history homework went straight to Dad for advice, while English, foreign languages and the arts were Mum's domain. My mother made sure that I did enough work to get good grades in these 'non-scientific' subjects. This meant that I studied English Literature in grade 11. English Literature was generally a grade 12 course, but I was able to do it a year earlier than usual (this meant that I could just do science and maths in grade 12 and get a really good score). After 6 weeks of school, we were told that Emily Bronte's *Wuthering Heights* would be one of the books we were to study that year. This foray into the arts had just gotten real! For a kid who liked action movies and falling out of trees, this was a nightmare. It is the most boring book. I tried to read it but after the first two chapters, I realised that I wasn't going to make it. So, I responsibly took the only reasonable course of action available to me: I changed schools.

The new college made me read better stuff like *Schindler's List*. But in term four, Thomas Hardy demanded my attention with *Far from the Madding Crowd*. This is another English Literature "classic" (which is code for boring). Even though the character development is exquisite, with a complex interplay of themes that paint a masterpiece of imagery, it all happens so slowly that I'd fall asleep reading it. The power of this book to make me sleep was exceptional. Once it did it to me while I was lying on a wooden bench with the sun in my eyes! What was I going to do? There weren't many schools left in my local area. I had to be flexible in my approach and try something new. I listened to the discussions in class and took good notes, which allowed me to write my essay about the book without actually reading it. Yep, you read that right. I passed English Literature (in the age before the internet was invented) without reading the text. At the time, I didn't recognise this as my greatest educational achievement, but literature taught me that smart-lazy (if used for good and not for evil) is a legitimate management technique.

I finished school with good grades and applied to the university that was situated less than 1km from my parents' house. That was an easy choice. I was accepted into the engineering faculty. Well, that is the short story; the long story is that I was first accepted into science because I thought that I might like to be a teacher. (I now know that I don't want to be a teacher). Instead of turning up, I took the year off. Towards the end of the "gap" year, I rang the university to say that I was ready to start. To my shock, the lady on the phone told me that I couldn't just start since my place had been for the previous year, and I would have to re-apply. (Incidentally, the university fees increased about 200% during my gap year with a change in the student loan scheme. Bummer.) During the re-application process, I made the pivotal decision to apply for engineering rather than science. This ended my teaching prospects instantly and led me on the long path towards the gaping maw of *the corporation*.

First year engineering was an interesting experience. Sort of like *The Hunger Games*, where only the fittest survived. You had to either be clever or able to motivate others. If you could motivate others to share their assignment answers, you didn't have to work out the answers yourself. Of course, sound judgement was required when deciding who to motivate... ideally someone clever. For calculus (difficult maths), we were required to submit an assignment every Monday morning, every week of the semester. I arrived early on the first Monday with a completed assignment (having worked pretty hard to complete it during the weekend). I made a decision to invest the assignment answers in as many people as I could that morning. I wasn't very judicious at all when distributing my assignment to all and sundry. In the weeks that followed, the returns on that investment were realised as I generally completed my calculus assignments on the Monday morning before class with the assistance of various parties who were still grateful for the assistance rendered to them in week 1.

There is an important point to be made here... my fellow students were all people. People with their own families, their own backgrounds, their own reasons for choosing engineering and their own reasons for not completing the assignment on the weekend. Looking back, I think that the most important thing I learned at university was how to get hold of stuff to use so that I didn't have to work it out myself. I can say with certainty that I have been effectively finding things to copy ever since.

People progressively dropped out of first year engineering. Maybe they discovered that they didn't like maths and physics, or maybe other opportunities opened up in their lives (probably the former, I think it's rare for 19-year-olds to have compelling opportunities open up for them). I do remember one chick at uni who was staggeringly clever – academically. She regularly scored the best marks and picked up concepts very quickly. It perplexed me that she wanted to design aeroplanes but had chosen civil engineering. A decade later, she was a singer/songwriter (she reminded me of Jewel). I guess working for *the corporation* wasn't all she had hoped it would be. Maybe if she had done aeronautical or mechanical engineering and had got a job in aeroplane design, she might have lasted longer. I heard her perform once... my assessment was that she was as good at singing and song writing as she was at learning civil engineering.

Most of the people who were going to drop out of the course did so in first year. I think that it was fundamentally kind of the university to make first year difficult enough to weed out those who couldn't finish the course, so that they didn't waste four years discovering it. By the end of the year, we had formed friendship groups. The group that I belonged to was really a partnership. It involved me and another dude called Jordan Spear. We were good together. We pushed each other academically and we pushed the boundaries of procrastination. And, we were both into sports. This partnership stood the test of time. Well, it lasted the whole four years of university. In retrospect, although neither Spearsy or I recognised it, we were two halves of a marriage of convenience – evidenced by the fact that in decades since we left university we have spoken only once (we ran into each other in the food line at the cricket).

In third year Fluid Dynamics, we had a lecturer who was irreverently referred to as Professor Three-Tone-Bone due to his wardrobe that seemed to entirely comprise brown items. He was clearly an engineer. Old school. It was perplexing why he was lecturing at university since he didn't like students. The subject he taught us was difficult, with lots of complex maths. I have thankfully never had to use the subject matter professionally. Professor Three-Tone-Bone would write about four blackboards of notes per lecture, and we had to scramble to keep up. Most of the course was about re-writing complicated mathematical proofs into our notebooks. These copied proofs would then be used in exam study, principally for rote learning. One lecture, while deep into blackboard number three, there appeared to be an incongruity in the maths. I put my hand up and caught his attention as he swapped smoking chalk sticks.

"Um, excuse me Professor, could you explain how you go from line 34 to line 35? I just can't see how the maths works." He looked at the almost full blackboard then turned to the student audience saying: "Well, it's complex algebra! I don't know how you do your complex algebra!" He then immediately went back to writing out the subsequent lines of the proof. Smackdown! My young but well-formed ego took a hit, and I sat in my seat looking at lines 34 and 35 to try to understand why I was so dumb that I couldn't make the mathematical link.

After a few minutes, I made the decision to advance my education at the expense of my reputation. I again got the attention of Professor Three-Tone-Bone and pleaded that he explain in detail the logical progression from line 34 to line 35 as I just couldn't follow it. He huffed and puffed and begrudgingly set about writing up about five lines of maths that made the detailed link between the two lines. As it happened there was an error in the proof. It originated in a + sign mixed up with a – sign, and the flow on of this within the algebra meant line 35 was in fact wrong. Wrong! I was right to question this anomaly even though the barrage of academic scorn was the price of exposing the error. I felt vindicated although I also felt a little bad for Professor Three-Tone-Bone, publicly embarrassing himself like that. I didn't want to make a big thing of it or anything – I just wanted to sit smugly and let the whole sordid affair be put behind us. Not so for Spearsy. He bellowed from the back of the room: "I don't know how you do your complex algebra! I … get it wrong!"

The whole class laughed out loud. Professor Three-Tone-Bone appeared not to hear anything and just got back to finishing the proof and then promptly left the lecture theatre early. In retrospect, I think this situation could have been handled better. Like Spearsy, I too had a tendency, when arguing, to completely defeat my opponent to the point of removing any scrap of dignity that he or she retained. This was no doubt developed by playing contact sport at school. My later experiences in *the corporation* taught me that winning in such domineering fashion didn't always lead to successful outcomes. There are times when you need to allow the other party to save face (even if they are your subordinates). This leads to loyalty rather than scorn.

The strangest course I did at uni was 2nd year electrical engineering. Electrical stuff is crazy; the maths is non-intuitive, and you often have to divide by $\sqrt{2}$ (I still don't know what this actually means). This was a key part of my decision to do mechanical engineering. The university where I studied offered three kinds of engineering: civil, electrical and mechanical. Basically, civil was boring with no moving parts (in fact, any bits of civil engineering that move constitute an epic fail). Electrical, as previously noted, is crazy. So, I pursued the meat in the sandwich – mechanical. As a mechanical engineering student, I had to do a course in second year on machines (motors) and transformers (not the fold-up-robot kind, but the device-that-changes-electrical-voltages kind). The lecturer told us that he was from New South Wales but we could tell from his look and accent that he was actually from India. His accent was so thick that we couldn't understand what he taught. To make matters worse, unlike most of the lecturers who taught by making you copy stuff from the blackboard (white boards had just been invented and hadn't got to our university yet), this guy taught by delivering a narrative while walking the room. He was probably an engaging speaker – he certainly worked the room. The only problem was that no-one understood a word he said. I remember him throwing a question to the audience, which was met with the silence of non-comprehension (of either the question or the underlying theory). The silence was total. He then singled out a guy and pointed at him demanding an answer. The guy (who was a 20-year-old engineering student, i.e. a smart arse) raised his shoulders, gave a gentle shake of his head and suggested "Seven?" No one knew what the answer was, but it wasn't seven. It probably wasn't even a number.

The problem was serious; we were halfway through the semester, and despite turning up to all the classes, we had learned nothing! The electrical guys needed this subject to progress into third year. We mechanical students, although not destined for even crazier electrical subjects, needed to get 45% in order to not have to repeat it. We put together a party of concerned student leaders and approached the Dean of Engineering to voice our concerns. The Dean listened to us and suggested that if he established a weekly tutorial for this subject, we would have an alternative path to enlightenment with an English-speaking tutor. We were saved! The next week we attended the first tutorial, excited for what the future held. The tutor, a masters electrical engineering student, was literate and understandable. She handed out a sheet with a number of problems on it for us to solve. Her instructions were for us to get started, and she would help us when we could not progress any further.

The seven guy piped up: "You don't understand; we don't know anything! We can't progress these problems at all, because we haven't learnt a single thing all semester!" She thought he was joking and wasn't going to get pushed around by a bunch of lazy second year students. She refused to help unless we 'had a go' first. A lecturer we couldn't understand and a tutor who refused to teach.

Lovely. Eventually, we convinced her to teach us something which turned out to be just enough to scrape through passes all round on the unspoken understanding that we would never attempt an electrical engineering calculation professionally.

Some of the courses we did were really hard and everyone knew it. For example, 'Maths for Engineers' was rough. However, some other subjects were only slightly more advanced than primary school level. One was called 'Accounting for Engineers'. It was taught on Tuesday nights by an accountant, not a lecturer. We assumed that he wasn't one of the best accountants since he spent his Tuesday nights with us. Presumably to earn extra money. I'm pretty sure it wasn't for our witty company.

The accountant started the course by asking who had a dad who was an accountant. A number of hands went up. One guy said out loud that his father was an auditor, and asked if that counted. The accountant took the question and smiled. "An auditor" he said as he leaned back onto the bench at the front of the theatre. "Does anyone know what an auditor does?" There was no reply. "Auditors are the modern-day equivalent of the people who ran around on the medieval battlefield after the battle was won or lost, bayonetting the wounded." None of us wanted to be an auditor. Or an accountant for that matter. No-one pointed out that bayonets weren't yet invented in the Middle Ages.

◊◊◊

Spearsy and I had a system. We would meet on the first day of swot vac (the holiday period before exams when you didn't have to go to lectures but were left alone to study) and we would plan. This involved setting aside which days to study for which subjects. We generally had five exams per semester and they were never evenly spaced out, so sometimes we had to study for the last exam first when we had extra days. We would allow about four days per subject; this seemed to be sufficient. We would check out the previous five years of exams, which typically followed a very regular pattern. There may have been seven questions in a thermodynamics exam of which you had to answer any five. Since we had the same lecturer who had set the last five exams, we considered that the patterns within those last five years were worth exploiting. As it turned out, the first six questions were essentially the same every year and then there was a 'wild card' question for number seven each year. So, we concluded that all we needed was to learn how to answer questions one to four in each of the previous five exams. Four questions – one day each – four days required to study for this subject. Ah, but that still leaves the fifth question that hadn't been prepared for. Well, we weren't too concerned about that since we only needed 50% to pass, and just by relying on natural brilliance, we were sure that we could pick up a few bonus points from any one of the last three questions.

A typical swot vac day involved meeting at either Spearsy's house at 9am sharp. We would then spend the next hour studying hard. We would take a half hour break at 10am till 10:30 to give our brains a rest. At 10:30 we'd get back into it till about 11:30 when we'd break for an early lunch. After lunch we would typically go and play basketball or footy for a few hours before knocking off for the day. We thought it was important to nurture the body as well as the mind.

Exams are a stressful and unnatural experience. You have three hours to prove your worth with no access to outside resources. This never happens in the workplace. Once you start work, you rarely get asked to prove yourself in any way, and on the rare occasions when you have a challenge, you always have access to resources (books, the internet, and my favourite: other people).

My worst result was a terminating pass in fourth year Gas Dynamics. I remember opening my results in mid-year to find among my scores a terminating pass. I was confused and disappointed. I was the guy who got 51% without trying (usually more like 63% without trying at all). With four days of study with Spearsy, I could better this to somewhere between 70% and 85%, which was right where I liked it… the sweet spot that maximised the score while minimising the effort. The 80/20 rule in action (Pareto principle, look it up). I booked a meeting with the lecturer to check that he had not mixed up my paper with someone else or maybe just made a mistake when marking. He showed me the paper and the only mistakes were in my answers. In fact, I was lucky to get the 45% terminating pass and not the 44% fail. I quizzed the lecturer on the impact of a terminating pass, and what I had to do to rectify the situation. His advice was that if I wanted to re-sit the exam, I could try for a bona fide pass of >50%. This would involve more study, which I wasn't terribly keen on, and I'd be on my own. Since Spearsy didn't get such a bad score, he'd be spending his holiday playing basketball or footy all day – not just in the afternoons. Upon further questioning, the lecturer advised that if I did nothing, the terminating pass was just that, a pass for the subject itself but not a 'qualification' to start the next course in the series. I asked if there were any follow-on subjects from Gas Dynamics. He said that as this was the 4th (and final) year, there were no follow-on subjects ever again. At that moment, I made a compromise in favour of my laziness and to the detriment of my ego (and academic record). There was a common saying at university: "51% wasted effort, 49% wasted year". But this experience had shown me that in this case, it took a lot less than 49% to waste the year.

All in all, the study technique employed by Spearsy and me proved successful for both of us as we completed a Bachelor of Mechanical Engineering with Honours - second class uppers.

I was now extremely educated and ready to find my place at *the corporation*. There were many to pick from. Most of fourth year uni was spent applying for jobs. I was interviewed by four *corporations*. Twice I was flown around the country to attend graduate engineer job interviews. The first day was aptitude testing; not just knowledge or pattern recognition but also personality testing. I remember one test where we were presented with a series of four word (or phrase) collections, and we had to select which word suited us best. For example: [flexible], [always right], [pulling rank], [strawberries]. I circled flexible, as it seemed the best answer regardless of how flexible I actually was. It seemed that the assessors agreed and I was put through to the second day of *The Hunger Games*. The main thing on the second day was being put in teams (I was in a team with Spearsy and six other kids from around the country) and we were each given a briefing pack (with not enough information) and told to work together to solve a problem without much time. We were seated in a circle with a larger circle of four assessors outside our circle. Not enough information and not enough time…I loved it! Immediately I started asking people questions and assigning them to find pieces of information in the briefing packs. After ten minutes or so, the exercise was brought to a halt by the assessors. That was lucky, since I was just about to run out of steam and was quickly coming to the conclusion that we were never going to be able to complete the task. I was through to the next round! Spearsy said that it wasn't surprising considering the way that I had commandeered the discussion during that exercise.

I had an inkling that I was a good leader at this point. In fact, I got that impression much earlier when I was made captain of my school rugby team. Rugby is a fantastic game. Unlike other sports which are mainly skills based, rugby is fundamentally toughness based. Even if you're not very skilful, you can win by having the courage to stand your ground and never take a backward step. This is epitomised by the legend of Buck Shelford, the All Black lock (and captain) who had his scrotum torn open during a ruck and demanded that the physio sew it up on the sideline so he could play the rest of the game. You have to be tough to be good at rugby. I found that I could get through to the other players on my team and draw out the best in them. It was about the right messages at the right time to keep them focussed on the goal at hand – motivating my team mates to be the best (toughest) that they could be while on the field. I realise that I have taken the opportunities wherever I could to get the best out of the people around me ever since.

Carston did second or third year with us. I say this because he started his nominally four year engineering degree at least a couple of years before us. And he finished at least a couple of years after us. He kept failing subjects. But full of never-ending determination, he never gave up. He just kept repeating subjects until he had studied every subject at least twice and scraped together passes for each of them. Carston typically sat at the very back of the lecture and knowing that he would probably be back next year it wasn't that important to listen to the entire lecture, but often thought of funny things to share with those sitting near him.

Professor Milenko was a Polish immigrant who lectured civil engineering subjects (rocks and concrete). He was an uncompromising, no-nonsense lecturer who probably hadn't laughed since before World War II. It was the last lecture of the year and for us mechanical students the last scheduled subject of civil engineering…ever.

"Carston!" Professor Milenko looked up from his blackboard to the back of the room. "Carssston, you think this is ffffunny?!" He seemed to linger on his f's and s's. Carston just smiled back. He knew Professor Milenko better than the rest of us since he had spent much more time with him.

"I have ssseen your marks, Carssston. That is ffffunny!!"

Soon after he dismissed the class and we filed out saying thank you to Professor Milenko as he stood still and nodded slightly and slowly without smiling. When Carston said thanks, Professor Milenko replied: "Sssee you nexsst year."

Big Arse was the professor who delivered the last lecture of fourth year engineering. That wasn't his real name, but it was an apt physical description. Like all the professors, Big Arse lived in the magical world of academia. (It's kind of like a castle where people with multiple degrees live together and think about complicated things. All the real work occurs outside the castle where the people with only single degrees do stuff. The university is like a lean-to, sitting against the side of the castle of academia. It acts like an air-lock, where both academics and students can mix, but neither group stays.) Every day for about 30 years, Big Arse had lowered the drawbridge, and, as though doing the "hokey pokey", put one large foot out to engage with us students. He taught us about beams. He taught us about the different standards that we can use to design beams. He taught us that bridges are beams. He also taught us about columns.

But he saved his most important lesson for last:

"Well, this is the last lecture for you guys. In a few weeks, you'll be out in the workforce. Don't think for a minute that any of you are going to change the world. The best thing you can do when you start your first job is to listen. And try to learn as much as you can from the people around you about how to actually do engineering, not just the calculations that we've taught you here. If you ever start thinking that you know best and have the urge to change things... don't. Just sit down and wait for it to pass.

"I'm just like you guys you know. I get the urge to go for a run. But do you think that I act on it? No. I have a cup of tea and a biscuit and wait for the urge to pass. Good luck with your exams."

I passed my final exams which meant that I was allowed to leave the university. I was ready to fulfil my destiny as a cog in the machine that is *the corporation*. I can safely say that I forgot 80% of what I was taught at university in the first year out. The remaining 20% took a little longer.

DESIGN (MY FIRST ~~LOVE~~ JOB)

I applied for a holiday job at *the corporation*. I was interviewed by a Team Leader named Pedro. He was a tall guy with a genuinely nice nature who sat on a strange desk chair that was more like a bicycle seat. He was also an avid cyclist. He said that they had a computer program that calculated the dynamic level in surge towers. It was written in QBasic and they wanted me to spend six weeks translating it into Visual Basic. During this interview I had a defining moment when I realised that I was out of my depth, and all the university training that I'd received clearly wasn't enough to prepare me for the workplace. I couldn't in good conscience lie to this obviously good man and take the job knowing that I didn't know how to do the task. I decided that the truth was the best policy. I said that although I had done a small amount of programming in a language called C++ (which is a silly name because it sounds more like an exam mark than a computer language) during my mechanical engineering degree, I had never used QBasic or Visual Basic. I politely backed out of the interview by stating that I didn't believe that I was the right person for the job and I even offered to ring around the computer engineering guys I knew to find one who was available. I was disappointed that this opportunity had turned out to be a dud, as I was so close to starting with *the corporation*. Pedro looked at me with a smile and said: "You'll be fine, start on Monday."

I was taken aback but immensely relieved. After my admission of incompetence during the interview, if Pedro was going to pay me to take on this task in full knowledge that I couldn't do it, then when the inevitable failure would come, he could only blame himself. And I would have been paid for the experience. I accepted and started the next Monday. Back then I didn't know that you basically make up jobs for engineering students to do on vacation. It's a bit like the jobs you get kids to do in the kitchen that aren't really necessary but it keeps them occupied.

On my first day, Pedro gave me a book that was titled: *Teach Yourself Visual Basic in 60 Days*. He said that I didn't have 60 days as my holiday job was only five weeks – so 30 work days, and maybe I should read two chapters a day instead of one.

As it turned out, Visual Basic was reasonably intuitive and it took me two weeks to translate the program. The guys in the team were a great bunch of people and during the next month, as I was sniffing around for more work, I helped them out on all sorts of calculations and other engineering things. A year later, my first graduate job was in the same *corporation*, in the same team. I even sat at the same desk.

I had accepted a job at *the corporation*. It was a state-owned power utility – it was a big fish in a small pond, and I was lucky to be there. I was a graduate mechanical engineer working in the design department. And loving it. I sat next to a Senior Engineer named Bill. He was a big guy, over 6ft, with a black beard. An old school engineer, he looked like he would have been just as comfortable building a footbridge (over a fast-flowing stream using only an axe and bailing twine in the snow) as he was doing spreadsheets at his computer. His wife was a dentist who was only working part-time in order to look after their young children. He taught me a lot about design and drawings and calculations and other boring yet important stuff. I remember a time when our new boss came over and said to Bill: "What if I cancelled your holidays next week because of how busy we are?" Like most of my bosses over the years, he was a bit of an idiot. Management positions seem to attract idiots for some reason. Maybe this is because competent people are content to do real work. It took me years to realise my potential in the ranks of management.

Bill stood up from his chair, towering over our boss and slowly said: "I can go home right now if you want." Our boss quickly backed out of the metaphorical corner that he had been put in, saying that he was just considering the options and not to worry about it and that he was late for something and had to leave. I was amazed. I thought to myself: I want to be like that. Standing my ground, no matter what. Just like Buck Shelford, Bill wouldn't let himself get pushed around by the whims of *the corporation*.

I discovered that the key to being in control was not having large debts. Debt meant that I would not be able to stand up to *the corporation*. Debt meant that if I was ever in Bill's situation, I would have to respond more meekly with something like: "Yes Master, and how may I serve you better?" I recognised that I was somewhat hot-headed and needed to control my debt in order to be in a position to be like Bill if the need arose. And the need may arise at any time, and without warning. *The corporation* can be a fickle beast.

This was the first time that I came across the corporate phenomenon that is timesheets. Some jobs need timesheets and some jobs do not. One of your career aims should be to attain a job where you don't have to fill in timesheets (management jobs). Timesheets are essentially a mechanism to allocate the cost of your time to various cost codes. (If you are unfortunate enough to need to fill in timesheets at your job, know that there isn't any reliable way that management can use to check what you put on your timesheet. If there was, they would use that system to fill in your timesheet in the first place.) I'm told by lawyer friends (acquaintances really, after all they are lawyers) that the normal practice in a law firm is to allocate your day into six-minute intervals. We'll call this a 'time unit' here. Ten time-units to an hour. Eight hours in a day. This gives eighty time-units a day to be charged to clients. Timesheets are the answer to knowing which clients to charge which time-units to. It's hard to imagine achieving anything in a six-minute period. Even walking to the kitchen and making yourself a cup of instant coffee and returning to your desk might spill over multiple time-units. In the engineering world we took a much more mature approach and allocated our time in fifteen minute periods. This "engineering-time-unit" was long enough to fully contain a trip to the kitchen or toilet.

I understood the importance of accurate timesheets and was diligent in making sure that mine were filled in correctly. The governing rules of timesheets were not always absolutely black and white. Some discretionary decision making was required. I had been in an aeroplane en route to have a meeting with client A. While on the plane I undertook some work (read a document) for client B. Who did I charge the time-units to? It was only reasonable that Client A paid for the travel time required to advance his project. But at the same time, Client B should be paying for the time spent on undertaking his work. Who should be charged? Maybe I should charge whichever one I liked the least (this was a tight contest). Maybe I should charge them both and then take time off on Friday to make up for my extra charged time-units. I needed help, so I asked Bill.

Bill stared at me as I explained my quandary. When I had finished, he simply said: "I think that you're taking this far too seriously." I charged them both.

The corporation gave out corporate credit cards to any employee who filled out the right form, so I got one. (37% of all professional time is spent filling in forms. The only way to avoid it is to get promoted into a very important role. Then you can cut down on your personal form-filling time because you would have a dedicated secretary who would fill them in for you. This would free up your time to be spent in decision making or golf. Secretaries spend 76% of their time filling in forms.) It felt good. I could pay for minor things like accommodation, petrol for hire cars or nuts and bolts without having to engage with the procurement department (engaging with the procurement department should be avoided wherever possible. This rule applies to every organisation). The only downfall was that you had to fill in a form for each transaction (and provide the receipt). This deterrent was made worse because your boss had to sign each of your transaction forms. Luckily, your boss invariably had many forms to sign and wouldn't have the time (or desire) to scrutinise any single transaction too closely. Each month the department secretary, Janet, came around hassling everyone to get our transaction forms submitted into the system. I didn't like filling in forms, and each month I waited until Janet started the nagging before I would think about undertaking this menial task. One month, Janet came around and told me that it was very important that I completed my forms. I replied that it seemed to be more important for her than it was for me. After all, I already had the goods or services from the transactions. Bill giggled. Janet did not.

At the bottom of the form, there was a section entitled 'Tax Status' with a blank box to be filled in by the credit card holder. This was one of the few times that I was truly stumped. There were no instructions on the form as to what this meant. There was nothing in my extensive education and training that shed any light on what I should write to describe the tax status of my purchase. I had nothing. So, I did the only reasonable thing to do. I ignored the task for as long as possible.

Then Janet started her monthly nag. So, I went to ask Pedro what to do. Team Leaders know these sorts of things. He examined the form and listened to my predicament. He sagely agreed that this was a problem then turned and started ferreting through his filing drawer. I was hoping that he would be able to shed some light on the problem. Maybe point out a flaw in my thinking or a new piece of information that cleared up the whole mess. I wondered how *the corporation* could function with such a gaping hole in the systems. Pedro surfaced from his filing drawer with a sheet of paper, saying "Aha! I have the answer." The form was a list of tax codes with descriptions. Some related to capital expenditure, some to operational expenditure (this is an important distinction for accountants. It isn't very important for engineers). We read the codes and descriptions together. Although I now had access to a list of possible tax status codes that I could use, I was no closer to being able to select the correct one. Pedro agreed it was difficult. I asked straight out what code I should use. He replied: "I presume that if it was important, the accountants would come and explain what the codes really mean. I just put GSTV on all my purchases and no-one ever complains." From that moment, I used the code GSTV for every transaction I ever made. I still don't know what it means. I was happy. Pedro was happy. Most importantly, Janet was happy. The accountants may have been happy or they may not have been – maybe they corrected the code…silently.

There were lots of systems in *the corporation*. As an engineer, you basically had to learn how to do them all. So, as a Professional Engineer, as well as processing credit card receipts, you had to find your own drawings, log your own correspondence, file your own emails, interrogate the system to find project costs, and even make your own tea. In days gone by, there used to be a small number of people who were specialists at all of these systems (librarians, secretaries, cost controllers and tea ladies). Those specialists had a skill set that engineers lacked. Namely, an attention span of longer than five minutes and attention to detail. The specialists knew the systems and sure, you had to enlist their help when required, but they made things easier. Somewhere along the way, these 'support staff' specialists were exposed as 'overheads.' That is, people who are not directly connected to core business and are paid for out of company margin. (If you ever realise that you are an overhead, beware – you are at risk!). The support staff specialists were gradually removed from the company (and the payroll). This meant that:

1. the payroll was slightly less;
2. which in turn implies greater efficiency has been achieved;
3. all the systems still exist (after all our superior systems are what set us apart from our competitors!);
4. trainers were employed to train all the staff (except for the uber-important) on how to administer the systems themselves;
5. engineers administered all the systems themselves...poorly.

◊◊◊

I was encouraged to do further education. (This meant that I was assessed as having potential. All the senior managers had more education than just a Bachelor's Degree in Engineering. Or so I thought.) I was offered the opportunity to do a Masters of Business Administration. An MBA. Post nominals. With this qualification, I could stick those three letters after my name (right after the seven letters of 'BEng(Hons)' that I was already using). All it would take was two years of full-time study and about $10,000 a year in fees. The company would pay the cost. It sounded good, although I couldn't study full time because now, I had this thing called a full-time job (that took significant time out of my week). The solution that most potential high flyers used was to study part time (just one or two subjects at a time rather than four per semester). Sign me up, I was already learning my way into senior management.

The university allocated my first subject: marketing. Marketing is not like engineering. Engineering is sort of like applied science. What I mean by this is that what you're taught can be traced back to scientific observation or testing. In short, it is based on facts. Marketing is not. It's better described as popular opinion agreed upon by expensive marketing professionals who try to present it so it looks like science. In short, it is a black art with no way of telling if your effort (and expense) has been successful or not. (The usual way of determining if your marketing effort is successful is to measure how much you have spent on marketing professionals. If they haven't been reassuringly expensive, you might have to buy some more.) I found it difficult to wade through the marketing gumf, agreeing with little and understanding even less. But still, for the first half of the semester I did all the recommended reading (for the first time in my education history). I completed my first assignment. Thought I nailed it. And I failed (39%). My lowest mark ever (even worse than gas dynamics!). I thought that I deserved a commendation for doing all that reading.

I was so put out that I basically boycotted the course and didn't do any more reading for the rest of the semester. The second assignment came around and I completed it and handed it in. To my amazement, I passed (63%). This gave me an aggregate total of 52% for the assignments which were half of the marks for the course. So, all I needed to pass the subject was to score 48% or higher on the exam. I decided that I should use the strategy that was yielding superior results and didn't study for the exam. I scored 56% which translated to 54% for the course. Pass. Bam!

What a joke. I pulled out of the course as it seemed farcical that the only way I could pass was by not learning. I couldn't bring myself to continue to be part of this university revenue raising. (Many years later my dad told me that I really didn't need a university-crested piece of paper to convince prospective employers that they should let me administer their business because I had actual experience administering businesses. After all, education is the poor cousin of experience.)

At this point I discovered that the qualification was not what mattered but that you had some post-nominals… two or three collections of letters after your name on your business card that implied education, intelligence, success. Since in engineering organisations almost everyone has an engineering degree, it's what else you have that is the differentiator. The universities started using the post nominals BE for Bachelor of Engineering but this does leave some confusion with the other BE: Bachelor of Education. (Don't want to get those two mixed up!) I have always preferred using BEng(Hons) after my name; it is very descriptive although technically incorrect, but most importantly it is eight letters long, ten if you include the brackets.

We have stumbled upon the two rules of what I like to call *the credibility principle*:

1. The more letters you can stick after your name the better; and
2. It doesn't matter which letters they are.

I counsel against generating random letters as post nominals or claiming any sets without the corresponding qualifications (although either of these techniques may work better than you'd think). Do your homework – some letters cost significantly more (time and money) than others.

The Institute of Engineers was pushing a post nominal qualification called 'Certified Professional Engineer,' CPEng (they were trying to make it like the CPA for accountants). The problem is that old experienced engineers generally refuse to do anything to gain credibility since they have experience instead. So, the Institute sought to generate Certified Professional Engineers from the graduate ranks. This meant that it would only take 25 years for those 'grads' to become the 'old hands' and then we would certainly look down our noses at any young up-and-coming engineers who hadn't even done enough to 'become certified.'

When compared to the MBA (Cost: $20,000 and four-years' part-time study; Benefit: three letters), the CPEng is far superior. It can be achieved in less than three years (I did it in 18 months), costs about $500 for the assessment, and most importantly, you don't have to do any study. The nature of the certification process is you write up 'Career Episode Reports' or CERs. This is basically creative writing that explicitly claims specific 'elements' of the work life of a professional engineer. For example, an element may be: *establish budget*. Buried in your CER you would include the sentence, "I established the budget for the project." Not "the budget was established" or "I was given a budget by my boss" as these do nothing to prove that you have met the element. As long as you copy out the exact words of the elements in the report, then those learned engineers who do the assessment of your CERs can tick off that you have accomplished that element.

In all there were about 100 elements that you had to claim. I had been at it for almost one-and-a- half years and had completed six CERs and had achieved almost half of the elements. One evening I did a quick planning session and mapped out what projects I had worked on and how I could claim all the remaining elements. The planning revealed that I had actually done enough stuff to be able to plausibly claim all the remaining elements (without any bold-faced lies). The only thing holding me back was to write up the reports and submit them. I did an all-nighter that night and wrote up seven additional CERs. After my darling wife proofread them all, corrected the spelling and grammar and made them readable, I submitted them all together. One month later I was interviewed. And passed.

I was awarded the five post-nominal letters CPEng. It took 18 months of engineering work experience, culminating in one all night effort of creatively describing my experience to reflect me in the best light possible (a bit like this book really). It cost a paltry $500. That is $100 per letter (or maybe it's more like $120 per upper case and only $70 for the lower case ones).

Being certified is like the gift that keeps on giving; if you have a CPEng then you automatically qualify to be on the National Professional Engineers Register. All you have to do is pay an extra $200 every year and you get another four letters: NPER. Four capital letters for $200 is dirt cheap. $50 each and they're all capitals! This represents extraordinary post-nominal value. (I have never found any actual value in the NPER qualification.)

It was glorious. I had gotten my first post nominals only 18 months earlier when I had studied for four years to get a Bachelor of Engineering (with Honours), and for a grand total of $700 I had just picked up another nine letters in two separate groups. I immediately got my business cards reprinted:

Jack Martino, BEng(Hons), CPEng, NPER

17 letters (excluding brackets)! I was educated. I was intelligent. I was successful. My business cards said so, even though I was still trying to work out what engineers do. I can now reveal that these 17 letters have been sufficient to carry me through my whole career in engineering. I have never seen the need to add to them. After all, having achieved credibility, further letters would be wasted. And probably cost more. And there isn't any space on my business card.

◊◊◊

The corporation was good at self-promoting propaganda. I bought it hook, line and sinker. I worked for the best company in the world! With the most advanced systems!! And the best people!!! My natural cynicism did shine through at times. I remember when there was a management reshuffle which involved the head of the generation arm of the business swapping places with the head of the consulting arm of the business. The benefits of this swap were communicated with the employees personally by the CEO in groups in *the corporation's* onsite theatrette. He said that there was no downside to this move because "a good manager could manage anything." I said to my colleague that I wouldn't know because I don't think I've ever had one.

But the fact is, I was lucky to be there. I learned how to be an engineer, how to control draftsmen (with lots of red pen) and I was given great opportunities to work on interesting projects. I had an engaging argument with a Project Manager once when we were working on an Alliance project. Alliance contracting is fundamentally a concession that the principal and consultant are unable to control a contractor through the use of traditional construction contracts (normally because they lack:

o   sufficient project definition at the time of contracting;
o   the discipline to not make changes post agreement; and
o   the courage to enforce the contract.)

This Alliance agreement was a three-way contract with the principal, the contractor and the consultant designer, who all shared in the profit or loss generated by the project. It worked quite well and co-operatively for most of the job until the project ran out of money. At this point, all parties degenerated back into their standard operating modes: the contractor started generating claims for additional payments; the consultant started protecting his own position by blaming the contractor; and the principal tried to reject all claims from both parties.

The Project Manager had set up this alliance project with a decision-making protocol in which decisions were made by consensus. Not the normal way where the person who is accountable for the outcomes of the decision gets to make them (later in my career, for clarity, I started describing this concept as: decisions are made by the guy who gets sacked if it's wrong). And not the less effective but still workable majority rules principle. Consensus. Everyone has to agree. Needless to say, even as a graduate I thought that this was ridiculous. I said to him: "I am a graduate engineer, why is this company relying on me to make all the decisions?" He looked confused and said that he didn't think that I understood what consensus meant. I countered that it was he who didn't understand the term consensus. Despite being young, I was a stubborn bastard and I would develop firm opinions on everything that I cared about. Since we would only make decisions by consensus, we would be at an impasse (and the project at a standstill) until everyone else agreed with me. I thought that others with greater experience and presumably better decision-making skills should be tasked with accountability for decisions.

◊◊◊

After three blissful years at the best company in the world, I felt that I should no longer be a 'graduate' engineer but should be allowed to join my professional brothers and sisters and just be called engineer. After all, I had been a star and done everything asked of me and had achieved a CPEng qualification from The Institute of Engineers. (Incidentally, this qualification meant that I was qualified to sign off drawings commissioned by the state government of Queensland. Although I was the only engineer in the mechanical engineering group who had this qualification, it wasn't that important because, we didn't do work for the Queensland government.) So, although Pedro and Bill supported my application for an upgrade to engineer, there was a problem. For *the corporation* to classify me as an engineer I would have to be upgraded by two pay levels in a single year. The levels were worth about $1000 per year each. It was explained to me that, although this was not impossible, it was improbable and would require a compelling case to rise more than the standard one level in a year. *The corporation's* position was understandable when you consider that without rules there's chaos. And nobody wants chaos.

We were at an impasse. *The corporation* wanted me to smile and appreciate the gift of working for them. I wanted to rid myself of the derogatory graduate title. And I wanted to get paid more. So, I took the unencouraged step of applying for another job. I was interviewed and was offered a new position with the job title of Engineer and a $10,000 pay rise.

Now I was in a quandary. The job offer was better in every way than my current job. Except for the fact that it was 'outside.' How could I leave *the corporation*? After all I had practically been brought up in *the corporation*. For three and a half years, I had been looked after, nay nurtured by *the corporation*. How could I betray all that love and care that had been showered on me? After all, this was the best company in the world. Of course, these things were never actually said out loud, it was just implied, subtly. It could even have been subliminal messaging on our computer monitors while we were doing spreadsheets. It was a very difficult decision to leave my first job. I even went back to management and told them that I had a job offer outside but that my preference was to stay with them. If they could only bring themselves to raise me two levels to engineer status I would gladly stay. I knew that this would leave me $8000 worse off, but it would be worth it if I could stay.

I was emotional when management told me that they wouldn't raise me two levels. I was angry that I was being treated so unfairly by *the corporation* that I loved. And that she was pushing me away into the dark and cold. But my stubbornness trumped my other emotions, and I resigned. My letter of resignation probably read something like this:

> *Dear Corporation,*
> *There's something that I need to talk to you about. I don't want you to be alarmed. It's not you, it's me. I just think that we should spend some time apart for a while. It will give us the space we both need to grow....*

I assumed that after a couple of years outside, *the corporation* would realise what a mistake she had made in letting me go. She would ask me to apply for a job within her ranks… and I would willingly oblige.

When I look back at this time, it amazes me how attached I was to my first job. Since then, I have left eight different *corporations* and it has become easier every time (Once I even told a difficult boss: "I've resigned from classier places than this before"). Since then, I have managed many graduates, and I have always encouraged them to do whatever they can to ditch the graduate title as soon as possible. And after about three years, I have encouraged them to go out and find another job where they can learn different stuff. This is the advice that I wish I had received from my managers and mentors. There is a big world out there with many more *corporations*. Some better, some worse. All different. And so much to learn and so many different people to meet.

I was good at mechanical design. I enjoyed it. But once my eyes were opened to the world of construction, I realised that I would never go back.

CONSTRUCTION

My new job was working as a construction engineer on remote sites building power stations. I was working for a contractor. It is generally believed that contractors are ruthless, often referred to as 'hard money contractors.' They would operate with German efficiency, since any wastage would lead to reduced profits. People told me that it would be tough, especially coming from my previous job as a graduate engineering consultant in a quasi-government organisation. Long hours would be expected and high performance would be demanded. This didn't really eventuate – the new place did waste slightly less money than the old place, but really it was just a new bunch of people under a different name trying to work out how to do their jobs. If we were lucky, we would make big profits. Mostly we weren't lucky and just made moderate profits.

It was at this *corporation* that I first came into contact with a bonus scheme. The new *corporation* was populated by about ten engineers and managers (we would refer to ourselves as staff); about 200 tradesmen (we would refer to them as trades); a handful of office administrators (they were not referred to very often) and the obligatory accountant (we called him the CFO even though I think he was really just an accountant). Those of us on staff were considered officers of *the corporation*. We were paid for 40 hours a week, but were expected to work any reasonable hours required to discharge our duties. We were, however, able to get paid extra money at the end of the year based on things like:

o   how much money *the corporation* made that year (the higher the better);

o   if we could claim to have achieved everything set out for us in our individual performance plans for the year; and

o   how many trades we had injured that year (the lower the better).

It was great. At the end of the year, you would do a performance assessment with your boss. You would give yourself a self-assessment of at least 100% then tell him that you were brilliant, pretty much solely responsible for *the corporation's* profits and that *the corporation* would probably fail immediately without you. Your boss would then 'moderate' your score down to about 70% while carefully considering your outrageous claims. (If any of your claims were sound, then he would use those same claims in his own performance review). At the end of the day it was the score that counted and you always scored about 70%. Any less and your boss looked like he had a turkey working for him, which was bad since it implied that he selected a turkey during job interviews, which would call into question his decision making. Bosses are paid for making decisions so there was a lot on the line. Any more than 70% and it might look like you were so good that you could do your boss's job (I won't explain why that is also bad for your boss).

After the crazy dance we called performance reviews was over, you got a big pay the next month. Your final performance score was never issued to you, nor was the calculation of how much money this score translated into. You were just happy to have extra money, and you felt good for that month. The trades felt good most months since they got paid for every hour they worked and often went hunting or fishing on the weekends.

On my first day, I was briefed by the state manager, who told me that *the corporation* was wonderful, it was great to have me on board, and I had a bright future there. He gave me a mobile phone and a set of car keys, and told me that my project was to build a power station some 200km away. He also told me that as the construction engineer, my job was not to manage the project (we had a separate project manager for that), but to ensure that the generating machines we were constructing worked well. He also told me that if I needed any help in technical matters not to ask him or any of the other managers because they didn't know how the machines worked. With some nervous excitement, I drove to the construction site in my company car, with my company phone. This was so much better already than my last job. I started to wonder why it had been so hard to leave.

When I arrived, the project was one year into a two-year build. We were refurbishing two large machines. Both were identical, they had both been stripped back with all the mechanical and electrical items removed, bagged and tagged. There was a warehouse full of shelves containing bits of steel and copper in plastic bags with black texta identification markings. There were also new components coming from France and Switzerland. With the deconstruction complete, the job was now to put the machines back together using all the new bits and some of the old bits. It was chaos and I loved it. My eyes were opened to the wonders of construction. Building stuff was so much more fun than drawing things for other people to build. I was never going back to being a design engineer.

I quickly recognised that the supervisors were super. I had stepped into a functioning team of senior supervisors (old tradesmen) who, as the name suggests, supervised the trades to do the work. They also planned the work, bought all the tools and stuff that was needed, and knew how to fix everything. Thankfully, the engineer that I replaced was a good dude and had worked in this team for a long time. He introduced me to the supervisors before he left, and this helped me to be accepted into the supervisor's club. But as an engineer, I was seen by the supervisors as a bit of a necessary evil. I think it was because I had a degree.

The corporation generally built two power stations at a time. It employed two engineers, one electrical and one mechanical (me). This meant that each project could only have one engineer on site. Of course, no power station project is exclusively electrical or exclusively mechanical. All builds are a combination of:

o   civil bits – things that don't move, like dirt and concrete;
o   mechanical items – things that do move (also known as machines); and
o   electrical systems – things made mainly of copper.

This meant that the engineer had to deal with problems outside his area of expertise. Luckily for me, I had been working for only three years at this point, so I wasn't burdened with much expertise to constrain my thinking. It was a huge learning curve getting up to speed with the subjects that I had avoided at university.

Generally, the best way to learn about something you don't know on a construction site is to ask a supervisor. You do, however, have to make sure you ask the right supervisor the right question. For example, if I were to ask Tom, the welding supervisor, about generators, he would have looked blankly at me at best (at worst he would have loudly and impolitely pointed out my error in judgement in asking him about electrical stuff and then gone on to question my engineering ability). However, if I were to ask Tom about the difference between welding methods, he would gladly teach me everything he knew. I don't think he was often asked this kind of question by staff.

Unlike the last one, this *corporation* didn't issue corporate credit cards. Minor purchases could be made on your own personal credit card and then be reimbursed into your bank account. In order for reimbursement to occur, you had to fill in a form and provide the receipt. The credit card forms here were never done late. The *corporation* had inadvertently set up a system where the accounting was in real time, because the forms were always filled in on the same day as the transactions. No secretary nagged you to fill in your forms, because if you didn't, you were in effect subsidising the company the cost of the purchase. I say inadvertently because the system was actually set up to avoid a repeat of recent fraudulent activities where a secretary (in another city) had somehow embezzled a million dollars through corporate credit card transactions.

While on this site, we had a site office with room for six supervisors, an engineer (me), a secretary and a very experienced site manager (James, the only other dude on staff – since he was tall with long hair and a very vague resemblance to Chewbacca from Star Wars, he was known as the Wookie - and he hated it). One day the printer stopped printing. Even in the age of computers, this is a big issue on a construction site. It was going to take days to get it fixed. Recalling my briefing with the state manager months earlier, and firm in the knowledge that my job was to build the machines NOT manage the budget, I set out to buy my own printer. I rang around the local stationery shops to find one that would accept a *corporation* order. This *corporation* still used the old-fashioned triplicate order books. How this worked is that if you wanted to buy something, you just needed to find someone with an order book and take one of the orders. The next thing you needed was a supplier who would accept an order instead of money. It surprised me how many vendors would accept the order – you could buy nuts and bolts, tools, hotel rooms, even printers. I drove my company car out to the stationer and exchanged an order for a new printer. The orders were in triplicate (three sheets). You wrote on the top sheet (white) and the next two sheets through the magic of carbon paper were automatically filled in. The second sheet (pink) was sent to head office where the accountant would presumably use the information on the order to actually pay the vendor with real money. The third sheet (blue) was kept on site and filed. Although it might seem unnecessary, it is important that the site keeps accurate records in case head office were to lose their pink copy, or if you had to explain to the accountant why you bought something that you weren't allowed to.

When I returned to the site office with my brand-new printer, the Wookie site manager, smiled at me and said that I'd better ring the accountant and find out what number to charge it to. I said that my job wasn't to manage the budget – my job was to build the machines, and that being able to print in a timely manner was crucial. He smiled again and said, "Just ring the accountant".

(I learned later that the Wookie had had firsthand experience of the 'opportunities' that the hard copy order system presented. He had a 'signing authority' of $10,000. This meant that he was allowed to write purchase orders up to that value; there were other detailed company rules about the signing authority, but this was the most important one. When the time came to do the high voltage bus bar work [specialist electrical stuff that needs special training because it's easy to die with high voltage electricity – you don't even have to touch it, it can be deadly just being too close - like lightning], he had sat down with the two electrical supervisors. The two supervisors shook their heads, crossed their arms and said, "Not qualified, not touching it." This body language is typical of supervisors; they can be very expressive when they don't agree with you. The Wookie site manager pointed out that the work was quoted in the contract price and we were obliged to do it. But the supervisors were steadfast in their refusal. The Wookie then rang head office to point out the issue and was told in no uncertain terms to get it done and get it done now. Following the clear instructions from senior management, he called in a sub-contractor [who was both capable and qualified] to do the work. He wrote them an order for $180,000 and got them started straight away. Over the next week, the work got underway and the pink page of the order was put into the mailbag, driven the 200km to head office, delivered, unpacked and put in the accountant's in-tray for processing. I can only imagine his face when he found amongst the $25.47 for fifty M8 bolts and the $267 for tooling ('tooling' was one of our favourite purchase codes since tooling was considered necessary for any job but you could be quite light on the details of exactly which tool you had bought), he found a $180,000 order for unplanned, and essentially unauthorised, high-risk, sub-contract work. The Wookie was summonsed to head office to explain himself. Apparently, despite being true, 'because you told me to' was not an acceptable response to senior management. In the end, the $180K was more of a budget estimate than a fixed price, and the sub-contract final cost was $260,000. The Wookie was punished by having to spend a whole day with the accountant to learn how to fill out purchase orders.)

I rang the accountant to ask what cost code I should charge my new printer to. He told me that I wasn't allowed to make a purchase like that (it was within my signing authority but a printer was considered a minor asset or something, so there were special rules, blah, blah, blah.) I said that I now had the printer in my possession and that the rules appeared to be more important to him than they were to me. And if he didn't want to tell me which code to write on the blue and pink order pages then I could send them both to him and he could fill them in himself. He considered the issue for a moment and obviously decided that this was not as bad as the Wookie's protocol breach. Rather than instigating an inquisition, he gave me a cost code to use. Oh, I felt good, I was on top of the world. I was pretty much above the law, beyond reproach.

I clearly didn't have to follow the rules that applied to the supervisors and the site manager. And I had my own personal printer that I plugged directly into my computer. These were the good times. (About one year later, I was back in head office, in between projects - with no work to do. These were the bad times, and the balance of power had reversed. I remember asking the accountant if it was ok if I got a new pen from the stationery cupboard. He said yes.)

One day we realised that a major machine component needed to go back to the workshop to be re-machined. This was a problem. The item, the thrust block, was around one tonne of steel that had to run like a Swiss watch. The turbine and generator shaft (so all the moving parts) hung from the thrust block. We had put it on to the shaft and taken it off again. Twice. You had to heat it up with flame throwers, then lower it over the machine shaft and let it cool down to normal temperature when it gripped really tightly. At the same temperature, the hole in the thrust block was smaller than the size of the shaft: a 'shrinkfit.' The problem was that it just wasn't sitting square on the shaft, which misaligned the whole machine. So, the Wookie (in his element) hatched a plan to send the thrust block back to the workshop to get it re-machined to a tighter tolerance. The only complexity of this plan was that the workshop was in another state (transport by truck, ship and more truck). And the only other complexity was that the client didn't know what we wanted to do, and wouldn't agree to re-machining without extensive measurement and testing. Meanwhile, every day we waited was prolonging the inevitable re-machining and delaying our construction progress.

After making a phone call, the Wookie said that he could get the thrust block on the ship that evening as long as we could have it ready for pick up in two hours. It was now time for me to make my phone call. I rang our office and asked for availability of intercity flights. The secretary said that the last flight that day departed in 45 minutes. "Book me on!" I replied. I was in my car before the phone call was over, starting the 15-minute drive to the airport that I did in 10. I rang the Wookie from the car to confirm that the thrust block would be on the boat; I would meet it at the workshop the next day. The next phone call was to an old mate of mine named Strings (when

we met at university I assumed that he was nicknamed Strings
because he was so tall, but it turned out that he was known as Strings
because he had injured his hamstrings so many times playing
football) to deliver the great news that I would be in his city that
night: "Short notice but, let's go out! And could I stay at your house?
And could you pick me up from the airport?" Thankfully, he replied
with yes, yes and yes. He was a very good friend indeed.

I had my laptop computer and nothing else with me as I
boarded the plane. When I arrived, Strings was good enough to take
me to a shop, where I bought a toothbrush and a pair of underpants.
I spent 12 hours at the workshop (the thrust block arrived before I
did the next morning) before returning home. Ever since then, I've
always tried to minimise the stuff I take when travelling. My rule is:
when you return home, if there is a single item in your bag that you
haven't used, then you've failed!

◊◊◊

(My darling wife said that she couldn't make sense of the next
paragraph. I said that it really needs a diagram to illustrate it better.
She said that a diagram would certainly make it the most boring
paragraph in the whole book. I admit that it is a little technical. I've
tried to simplify it as much as possible, but leaving just enough detail
to lend plausibility to the story for any other hydro-engineers that
read it.)

Hydro-electric turbines are driven by water (*hydro*) to make
electricity (*electric*). The deepest part of the machine is where the
high-pressure water turns an 'impeller' (like a propeller but in water)
in the *turbine* that spins the generator producing electricity. The water
enters the turbine at high pressure and high speed and falls out the
bottom at low pressure and low speed having been stripped of its
energy. This is the miracle of hydro-electricity (the only dispatchable
renewable energy source – wind and solar are the poor, delinquent
cousins of hydro power). There was a problem with the painting on
the inside of the turbine 'Spiral Casing.' The Spiral Casing is like a
snail shell on its side. Kind of like a spiral (engineers are not terribly

creative when naming components). Imagine a long cone that is bent around on itself in a circle. The large end where the water enters was about two metres diameter (so you could stand up in this bit) and it gets progressively smaller to a pointy end as you go around the circle.

We had painted the inside of the Spiral about a year earlier, but it turned out that there was a problem with the temperature of the drying paint that left a microscopic film of moisture between the paint layers. This meant that the top layer was peeling off very prematurely. The upshot was that we had to go back in and repaint it (and take better care to control the temperature this time!)

Hydro power stations are very deep on the inside. From the outside they look like a normal building, but inside they go down a long way. The turbine floor is the bottom floor. It took a while just to walk down six flights of stairs to get to the turbine floor and longer to get back up. When the isolations were completed, and following the long climb down, the Spiral hatch was opened and immediately a dreadful stench emanated. It was, peaty (although not like scotch), fishy and decaying. We climbed back up the stairs to get advice from the power station operators who without going down to the turbine floor said that this was pretty normal, probably some algae or something, just leave it a day or two and the smell will clear. But the smell did not clear. Two days later it was as bad as ever. In fact, it was getting worse and the whole turbine floor was beginning to smell like an Asian fish market at the end of a sweaty day.

You couldn't see much from the hatch so, someone had to go down into the Spiral to investigate. I had been in the spiral a few times (it's a pretty eerie place designed for water not humans). And I was sort of excited to go in again to find out what was causing the smell. A brave young fitter accompanied me. We set up the gas detectors, establishing that the air in the spiral was safe to breathe, and carefully climbed in. Activating our torches we found that the whole spiral floor was covered with a thick layer of dead eels. And not just dead eels, but dead eels chopped up into about 40cm long pieces. Somehow a bunch of eels had gone through the machine, been chopped up by the turbine and we must have shut down the machine at just the right (or wrong) time to capture the pieces. At least we had found the cause of the stench. And it was disgusting. Thinking back on it, the only good news was that flies couldn't physically reach this place. Flies would have made the situation unbearable.

How to fix the problem? We could start the machine back up and wash them out, but it would take a couple of days to de-isolate everything, start the machine up, wash away the eel pieces, then start the shut-down procedure again and re-isolate. And there was no guarantee that there wouldn't be fresh eel pieces to contend with. A quicker solution would be to find a couple of guys willing to wade into the eel carcases and carefully shovel the pieces into the turbine impeller where they would fall into the tailrace (where the low energy water escapes the turbine). If only we could find a couple of guys willing to get back into the cold, hard, wet, slippery, dark, stinking, rotting-meat locker that was the Spiral…

About two hours later, after clearing the large part of the turbine, I found myself lying, stretched out, face down in the tight end of the spiral reaching forward with my shovel to scrape off the last eel bits, flipping them into the impeller. My faithful fitter was behind me far enough back where he could still stand up. He had a hose and was "helping" by squirting the water over me to my shovel tip which was trying to scoop-away the last pieces of eel. As the water and 'eel juice' flowed back down the Spiral 'floor' it oozed across my hands, face and neck, through the cuffs and neck of my water-proof overalls going down my arm, to my armpit and then all around my body. I could almost taste it. So much for the water-proofs! It was rank. I heard my fitter mate giggle a little as he asked:

"So, Jack, when you was at university and stuff, did you ever think you'd have a job doing this?"

We got the last of the eels out and I left the next crew to start drying the Spiral in preparation for the painters. I, on the other hand, went upstairs to our site office, got changed out of the stinking eel-juiced clothes and put them straight in the rubbish bin. Then I drove to my accommodation with the windows open. When I got there, I removed my newly basted fishmonger's clothes and put them in the outside rubbish bin. It took three hot soapy showers in a row to disguise the all-pervading scent of eel. I bought replacement clothes on the way back to work that afternoon. I wasn't sure whether I had earned some respect in the eyes of the trades that day or whether they just thought I was an idiot. After all, it was I who they all saw emerging from the spiral dripping in three-day old eel entrails.

Many months later after the smell and memory of the eels dulled, we were commissioning the new generators. We had built a state-of-the-art condition monitoring system. If it could be measured, we measured it. The old-fashioned way of doing condition monitoring was to have two horizontal accelerometers (that measure acceleration) on each of the three machine bearings, plus one on the top bearing measuring vertically. Seven measurements in total. This method had been used for decades and everyone who knew how these machines worked could interpret these readings reasonably well to diagnose the machine's performance.

The new way was to hang $1.5 million dollars' worth of sensors off every part of the machine and measure everything that could be measured: position, displacement, velocity, acceleration, temperature and probably operator fatigue. After all, everyone accepts that more data is better. So, the measurements were logged at the rate of six measurements every second over about 300 sensors. This gave a grand total of 108,000 data points every minute! More data is better. The French guys who did the designs for us said that it was a waste of time: machines of this size don't need much condition monitoring at all.

After commissioning but before handover (in that dangerous period when the contractor is still responsible for the machines but isn't running them), there was an unexpected trip. (This is not a surprise vacation – don't forget your toothbrush.) A trip is when a machine shuts itself down, uninitiated by the operator. It's considered very bad (although not as bad as the machine not shutting itself down and keeping on running until it destroys itself - just saying). As the degree-qualified mechanical engineer who supervised the construction of the machine, it fell to me to diagnose the cause of the problem.

Easy! We had big data on our side although even I didn't realise just how much data we had. I had the analyst send me all the condition monitoring data for the period ten minutes leading up to the trip and ten minutes after the trip. The data arrived on two DVDs. I loaded it all up on to my computer and graphed it all – 300 lines on a graph that showed the complete status of the machine every 167 milliseconds. I looked at it for a few minutes wondering how I would ever make head or tail of it. This could be the end of my career since I had no way of navigating through this problem to diagnose the machine fault. I didn't have the faintest idea how to interpret most of the measurements. Then it came to me to get rid of the ones that I didn't understand. So I started deleting lines of data sets. Slowly at first, I eliminated an obscure temperature reading that I didn't even know where the probe was installed. But pretty soon I was deleting whole groups of data sets in broad blows. I deleted all velocity measurements in a single key stroke. Eventually, I was down to the seven accelerometer readings that I understood, which gave me enough information to diagnose the problem.

Contrary to popular belief, more data is not better. Most of the data collected here was stored for no-one to ever look at again. And on the occasion that it was needed, almost all of the data was ignored. The French guys were right.

◊◊◊

A good week was one when you didn't have any contact with management from head office. The construction team working for the Wookie worked six days a week (with Sunday off because it was too expensive to pay double time for Sunday shifts). The guys from the client organisation who were seconded into the construction team worked a nine-day fortnight. (Monday to Friday on week one and then Monday to Thursday on week 2. I asked Reginald, the old crane driver how the nine-day fortnight worked. Did the whole team take every second Friday off or was it split so that half the team were there every Friday. He said that he had every Friday off, he just had to turn up for every second one.) I typically worked almost fifty hours between Monday and Thursday and then, drove home for a long weekend every weekend (sort of like Reginald). As I'd leave the supervisors office, I'd say good luck and to give me a call if they needed anything from me before Monday. After all I was only supposed to be working nominally forty-hour weeks.

Occasionally, the Project Manager (who was based at head office) would feel like a drive in the countryside and would come to site. He was called many things but I shall refer to him by his abridged job title here: PM. No-one really liked it when he was on site. They all felt that they were being watched. The whole site felt different. Maybe this is a natural phenomenon – the serfs don't feel comfortable with the king hanging around, pretending that he knows how to do their jobs. But the PM didn't do himself any favours since he was incompetent and a bully. This is a particularly dangerous combination. A classic 'seagull manager,' he would come to site, flap his arms around, squawk at people, crap on everyone then fly away.

I had seen him yell at people on a number of occasions, and my first reaction was to try to make sure that I was never in his firing line. But one afternoon, I got into his sights. It all happened rather quickly. Like a rabbit in the headlights, I didn't have time to escape. One minute I was busy at my computer (probably printing to my personal printer), when the PM came over and told me that I had cocked up the build of the generator stators twice, and I needed to fix it faster. He then hit me hard on the shoulder. His assertion was correct (we had twice failed in stacking the stator parts within acceptable limits and were now doing it for the third time, and since my job was to make sure that the machines were built well, it was indeed my accountability), but I was really surprised by this display of violence. I had never heard of someone in a professional workplace being physically bullied. Sure, sometimes it used to happen during apprenticeships, but even that had been largely stamped out a generation ago, I thought. Let me make the point that I was quite comfortable with this kind of violence in my leisure time (having played rugby for many years), but this display was workplace dominance. We had become pack animals and the big dog was making sure that all the puppies knew that he was the alpha. He needed to keep control by making sure that all people on the site understood his superiority. Especially the engineer. (I think I was chosen as a high value target because I was on staff. Bringing me to heel would send a particularly strong message to the hordes of trades who may have been planning an uprising.) I began to shake as my fight or flight instincts kicked in. My mind whirled as the adrenaline pumped through my system. I was not going to put up with being treated like this. I had never been dominated on the rugby field and I was damned if I was going to let it happen now. Would '*I-can-go-home-now-if-you-want*' Bill allow himself to be treated like this? I knew the answer to that – no! No, he would not. The world around me seemed to move in slow motion and I realised that I was trembling as I said in my firm but careful voice: "Don't hit me, it's… workplace violence." Could I have come up with a more geeky phrase? Not at short notice.

He looked down at me with an *are you challenging me, boy?* look. "I'll give you workplace violence!" he barked, loudly enough for the supervisors to hear. The threat hung in the air.

I knew how Bill would handle himself in this situation and my trembling was beginning to calm. I slowly stood up from my seat to look the PM in the eyes and simply said: "No you won't." I braced for impact. I had stood up to the big dog and now expected to fight to the death. Only one would survive. I actually expected to have a punch up in the office. I was pumped with energy, ready to go, just awaiting the PM to strike the first blow (which would technically be the second blow when it came to witness statements in court), and then it would be on. I'd probably lose, as he was bigger than me, and considering his senior position, he must have done this many times before.

He didn't hit me. Without a word, he just turned around and walked away. He then left the building and got into his car and drove back to head office. I was still trembling, rooted to the spot, surrounded by supervisors still sitting at their desks who had turned to witness the spectacle. There was silence for a while until we heard the PM's car ignition start, then the supervisors, one by one, swung back around to their desks. One of them got up and on his way to the toilet, gave me a pat on the back and said: "Well done." I had been accepted by the pack.

I sat back down at my desk. What was I thinking! As if anyone had fist fights at work. Well, professional boxers do, but we were not boxers and only borderline professionals. When the trembling finally subsided, I realised that I had learned to stand up for myself, just like Bill. I would not tolerate bullying of me or anyone else in my care. I had more power than I realised. It was important that I used this power for good and not evil. I rarely spoke to the PM again for the next two years that I worked in that job. He focussed his attention largely on the Wookie and the supervisors. If he could preserve his hold on them, maybe he could keep the illusion that his kingdom was intact.

Although I have witnessed workplace bullying many times since (and done everything I could to stamp it out), this was the only time when it involved physical violence. I still sometimes tremble when going into crucial conversations that might lower my standing within *the corporation* (this is a euphemism for 'I might upset one of the big bosses and he might sack me'). But I've found that being steadfast to my beliefs (normally expressed as my refusal to comply with idiotic or immoral directions) has never got me sacked. If anything, in a strange way, it has usually endeared me to my bosses. This was never my goal.

◊◊◊

Back at head office, we worked in a single open-plan office with 16 desks in groups of four. Each desk was partitioned off from the other three in the group with little plastic fences that were low enough that you could see over them, even while sitting down. Once upon a time, people were allowed to sit in their own offices. Then the offices were replaced with cubicles (which were sort of like mini offices without the top part of the walls), and then the cubicles became desks with only little fences to give the illusion of privacy. There was a lot of rhetoric about the benefits of "open-plan offices," largely focussed on superior communication. Apparently, it had nothing to do with open plan costing much less than building separate offices. Of course, there were still a few offices for the truly important people - presumably they didn't need to communicate better.

One of the old managers was a table thumper (as in, he thumped the table in meetings while yelling orders at people). He was never happy with me but it wasn't personal – he was never happy with anyone. I had no tolerance for this kind of behaviour, and I avoided him wherever possible. Most of the guys in our open-plan office worked for him, but I did not (by virtue of me having a degree). At times when we were forced into contact, I would revert to sarcasm. I had informed him that I was going on leave, and for a particular job, I had done all the preparation and was about to hand it over to one of his guys to perform while I was away. The next morning, I came into the office to find him leaning on the wall near my desk seemingly waiting for me. "What time do you call this?" he opened a conversation with me.

"Great to see you again, this fine sunny morning," I gave him my biggest smile.

"Don't you lie to me, Jack." I had never before this moment realised that he had a sense of humour. It was the only time he ever made me smile. He went on: "You can't just piss off on holidays and leave this job in the lurch, you know." It was a statement not a question and delivered loud enough for the whole office to hear. The old daddy lion was out to show the cubs that he was still the king of the jungle, and even the young engineer who didn't report to him would bend to his will. There was a certain sense of déjà vu for me.

"Sorry about the misunderstanding," I replied with a similarly loud voice, "but I wasn't asking your permission to go on leave, I was informing you that I was going on leave. I presumed that you'd like to know about it before I left." He smiled and retreated to his office without another word. He didn't get into the position he was in by fighting battles that he couldn't win. The guys in the gallery of desks also smiled but made sure that they were looking at their computer screens as he walked past.

◊◊◊

I got a call from one of *the corporation's* other workshops in regional New South Wales. The dude asked me if I could help him build a lathe (a machine that spins metal pieces that are then shaved into shapes). I was hesitant at first – after all, I had never designed or built a lathe before, or anything similar. Why ask me? But I flew out to visit him anyway, to see if I could help. As it turned out, I was a really good choice since I knew what a lathe was, and had seen them in operation in our workshop. I met with them at their workshop: an engineer and an old Croatian fitter/machinist (a guy who could operate lathes among other things).

The engineer explained that they wanted a mobile lathe to take to coal power stations in order to machine steam turbine rotors on site. This was advantageous since it eliminated road transport across the country by truck (which cost a bit, but more importantly, if the truck had an accident, then the turbine would be down for months or years). They explained what they wanted:

o   a mobile lathe;
o   that could be taken apart and stored in a 20 ft shipping container;
o   that could be assembled or disassembled in six hours; and
o   that could handle a 20 tonne piece.

"What makes you think that it's actually possible to build such a thing?" I asked the burning question. The old Croatian replied:

"I saw one once in Europe."

"What did it look like?"

"I'll show you the photo I took." Now we were getting somewhere. He pulled out a square polaroid photograph taken in the 1980s.

"Well, if other people have done it, then we can too," I supposed.

I took a photocopy of the photo and returned to my workshop with a lot to think about – after all, I didn't have any design software. I didn't really know what I was going to do. I worked on the concept with my draftsman for a day, and we did up a General

Arrangement drawing of what we thought would work. I also started talking to people I knew about all the stuff that I didn't know. Luckily, I had worked with lots of different people by this point and through this network, I got enough advice for me to make decisions about how to design the machine.

At the appointed time, I emailed the drawing to the engineer and rang him to talk it through. The sticking point was the cost. It was going to cost $190,000 to fabricate. He hesitated and said that he only had $150,000 in his budget. "I could not paint it; that would save about $10,000," I offered.

"Nah, go ahead, I'll take the kick up the arse if it comes!" This guy knew how to take risks. "One other thing, we need to use it on site in seven weeks on a job. We just signed a contract for on-site refurbishment, and we sort of told the client that we had one."

No pressure.

The next six weeks were hectic. My draftsman did detailed drawings of the components each day. The component drawings were then sent each afternoon to the steel cutters, who would deliver that day's steel two days later. So, the welders in the workshop were working off drawings that were done two days earlier. The only real hiccup was that the first four batches of steel came with each piece 6mm too small. This was because there was a new guy on the laser cutting machine at the steel cutters. The laser was 6mm thick and he set up the cut measurement to the centre of the beam rather than 3mm offset. So 3mm short on each side, i.e. 6mm short over the whole length. Every piece. The welders knew how to fix it by chocking 6mm strips into each weld. (It did lead to a contractual dispute with the steel cutters for the cost of the rework).

The component moved from the fabrication shop (welding) into heat treatment (baking the bits overnight in a really big walk-in oven to release any tension that the welding may have created) and then into the machine shop (for machining – basically making the bits exactly the right size and some surfaces smooth).

We did a trial assembly, sent some photos to the engineer and bought a 20-foot shipping container. Then we packed all the bits into the container and dispatched it to site on the other side of the country.

We waited with bated breath for news from the job, wondering if it would actually work or not. As it turned out, the lathe worked wonderfully well. After the job, the engineer told me that the lathe had a 17-day payback period. Most businesses are extremely happy with a 2-year payback period. Exceptional.

This seven-week project was the most exciting one I have ever worked on. It all happened so fast; I had to make solid but quick decisions every day. There was no time for second guesses and no room for mistakes. I loved it. I had learned to back my own judgement. And most of all, I had learned that I didn't need to be able to see all the way to success. The rush of not knowing it all, but needing to know just enough I found exhilarating. I just needed to be able to see far enough to take the next step. If I kept doing this for subsequent steps, eventually I would succeed. From this time on, I craved the jobs that I hadn't ever done before.

In seven weeks, I had become a lathe designer. I wrote a paper about it. It was really a puff-piece saying how good *the corporation* and I were. It was published in the local Engineers Australia magazine. Secretly, I hoped that there were other people out there who would realise that what they really wanted was a big mobile lathe and I could regain my hectic excitement all over again. But alas it panned out that Australia wasn't big enough for two large mobile lathes. In fact, it didn't really seem big enough for One. After the initial glorious project, I'm pretty sure that my lathe was never used again. It's been packed up, neatly stored in its shipping container for decades.

◊◊◊

The corporation had offices all over the world. The designers who contributed to the work we did were based in France, Switzerland and Spain. A big part of my job was to get designers in Europe to help solve construction issues in real time as we created them. Typically, I would be on the phone (my company mobile phone) at 4pm every afternoon. This was 8am in Europe. I would pace the long corridor while deep in correspondence. It became a ritual on site at 4pm: come and watch the engineer walk and talk. My favourites were the French. My mother is Italian and my primary contact in France was a guy named Jean-Pierre, whose mother was also Italian. We spoke Italian to each other to begin our phone calls. I loved this because it was a very clear message to the audience of supervisors who were listening that I was staff, not just because I had an engineering degree, but because of my advanced and rounded education – which included the arts (and at least one foreign language). Once we started talking technical, I had to revert to English.

Interestingly when the French guys came out to Australia, they seemed enamoured with two things. A particular seafood restaurant on the waterfront and checking out the road kill on the country roads to the power stations. If you drive out early in the morning, there would inevitably be a few dead small wallabies or similar on the road from the night before. The French guys had never seen a wallaby (which they described as a kangaroo not knowing that they're a lot bigger) so they made sure to observe the carcases on the road. Just thinking about it now, maybe we should have taken them to a wildlife park or zoo so that they could see living wallabies or even kangaroos.

It was recognised by my senior management that relationships were important. Crucially important. If I were to just send Jean-Pierre an email asking for resolution to a complex and urgent issue, the email would be added to his inbox along with similar requests from other projects around the world. To make matters worse, English was not his first language and there might have been elements of the description that were lost in translation. How did I get Jean-Pierre's attention to attend to my email as a priority? (most of my questions required resolution overnight so that the next day shift would not be held up). The answer is relationships. If I had a strong relationship with Jean-Pierre, he would see my name pop up in his inbox, and he would want to read it and help me.

However, it takes more than a phone call to become friends. It takes a number of face-to-face meetings. Because I'm fairly likeable, I can develop the foundation of a good professional relationship within about three face-to-face meetings (of course I couldn't do this with everyone but most people seemed to fit this pattern). I had met Jean-Pierre here in Australia a few times and since I liked him, I had invited him over to have dinner with my family. I remember this as Jean-Pierre went to great lengths to explain that 'coq-au-vin' needed to be made from an old rooster rather than a young chicken. And consequently, the dish that I had prepared, although very nice was not actually 'Coq-au-vin' but really just a chicken casserole. He did it delightfully politely even though he was French.

(This is a tip I learned from my dad. Eating meals with strangers is the best way to stop them from remaining strangers. Just because we work for the same company doesn't mean that we can't behave as friends. We might even become real friends and enjoy working together even more. It's not a new concept – human beings have been doing it since biblical times.) We had a pretty good relationship and my management saw the benefit of investing in it further. So, I got an all-expenses-paid, three-week holiday to Europe! *The corporation* called it a training exercise – we called it a junket.

I built relationships all over Europe. I built relationships with the Spanish office over midnight tapas in Barcelona. I built relationships with the French office by learning just how good Bordeaux reds can be. In Switzerland, we built relationships with fondue. This is a Swiss invention where you have a communal bowl of melted cheesy mixture (I was told that the exact recipe is a secret known only to Swiss men). You each have a pile of bread cubes and a skewer. You skewer the bread, dip it in the cheese mix and then eat it. It's frowned upon to drop your bread into the cheese. It's a brilliant novelty meal, and if you ever get a junket to Switzerland, get your work host to take you out for fondue – there's a great relationship in it for you both.

While on my Euro junket, I spent a weekend in Marseilles. I took a boat tour from the old port into the bay to visit the Château d'If, the island prison made famous as the place of incarceration and subsequent escape of the Count of Monte Cristo in Alexandre

Dumas' book. (It's interesting to note that this famous book was fiction and not a biography. The Count didn't really exist and certainly couldn't have tunnelled out of the Château with a spoon, since the whole island is one big rock. Dumas wouldn't have known this since he never visited the island. Maybe he just read about the place in other works of fiction. The Chateau d'If's most famous prisoner never actually existed. Interesting.)

There was a dude on the tour who was tall and blonde. I had a sense that he was an English speaker but I couldn't put my finger on why. I thought I had developed a very mild superpower. At the Château, I approached him and asked: "Where do you come from?"

"How do you know that I speak English?" he replied in a Pommy accent.

And then, in a cultural epiphany, it came to me: "You've got one side of your shirt tucked in and the other side untucked. No self-respecting Frenchman would do this. He'd either have both sides tucked in or both sides out."

"I live in London." I saw him later on, back in Marseilles, with his shirt tucked in on both sides.

I've been to Europe a number of times, and I always take a pair of track-pants. Every time I pack, I have a bold notion that I will wear track pants in France or Italy. This is some sort of Australian bogan pride coming out. I like trackies; I find them comfortable. When I'm in Europe, I know I will continue to find them comfortable and plan to wear them. Maybe just in the hotel, maybe out on the street. I am Australian after all.

However, when I get to Europe, I never wear them, not even in my hotel room. I can't bring myself to be the only man on the continent wearing track pants. With my darkish complexion, I find that as long as I wear jeans and keep my mouth shut, I can pass as French or Italian. This illusion disappears as soon as I speak.

It can be fun to locate the English-speaking foreigners in the population who are breaking the fashion code. Things to look for include:

o trackpants (as discussed above);

o polar fleece anything;
o shirt half tucked in (also discussed above); and
o running shoes without running.

Any of these tell-tale signs imply that the wearer can be safely approached and conversed with in English. He or she is necessarily from Australia, New Zealand, Canada, South Africa, England or the USA. (Note that unlike the colonial countries in the aforementioned list, the USA variety of English speaker is usually first identified by sound rather than by sight.)

◊◊◊

My direct boss (his name was Nick Stevenson but everyone called him Steevo) didn't understand the importance of relationships. He was an abrasive character who tended to bully people, and his actions said that he was happy to burn relationships for short term gains. He didn't care that most people didn't like him. Fair enough. But in a dazzlingly bright blind spot of his, he didn't realise that this was bad for him. It doesn't matter which company you work for, there are always some people that you need a good relationship with. As a minimum, you don't want to get on their wrong side, or your effectiveness will be severely hampered. This in turn could lead to the shame of less than 70% on your performance review. These key people are generally not considered very important, but they wield more power than you (or they for that matter) realise. They are: (a) the guy who fixes your computer; (b) the girl who books your flights; and (c) if the company has a store, the storeman.

Steevo treated the receptionists particularly badly (they also booked the flights). They pretty much hated him. I was in reception one day at about 4pm and witnessed one of the girls take a phone call. She ended the call by saying: "Leave it with me, I'll get straight on to it." I asked what it was about. She told me that Steevo was stuck in Darwin, his flight was delayed, and he wanted to get transferred on to another flight instead of waiting for half the night in Darwin Airport. With a smile, she stood up, gathered her bag and coat and went home for the day. I'm pretty sure Steevo was not smiling.

When the project was completed and we didn't have a power station to build, we were all relocated back to head office and awaited one of only two possible outcomes. Either *the corporation* won another job and we would all set off back into the field to build it. Or, there wouldn't be another project and we would all get made redundant. I couldn't stand the boredom of waiting so I got another job.

I took a job with a consulting engineer. Having already left one job, it was much easier the second time. Basically, I knew what to do. At that time, I was in the middle of a dispute with the HR department (Human Resources) because I had been sick for three days during recent holidays, and I wanted to change three days of leave into sick leave. Even though I had a doctor's certificate, I was told that this request could only be processed if the state manager approved it in writing. So, I booked a meeting with the state manager. At the appointed hour, I went into his office (the biggest office with the best view and the best desk – this is necessary to make sure that everyone knows that he is the most important) and announced that I had two things that we needed to discuss, and that it was important that we did so in the correct order. I had two pieces of paper in my hands. I tabled the first paper. It was my leave request form. I explained the situation. He listened to my logic, nodded and signed my leave request form. I then tabled the second piece of paper, my letter of resignation.

The Wookie resigned days after me but somehow left the organisation before me. Over sausage rolls at the morning tea on his last day (every company seems to do this; you get to eat sausage rolls when someone leaves), the state manager said that he wanted to thank the Wookie for his contribution over a number of years, and that *the corporation* greatly valued him (there was no mention of The Wookie's purchase order calamity). The Wookie responded by saying: "Well, you obviously don't value me that much, or I wouldn't be leaving." The room became suddenly cold. The sausage rolls, however, were still hot.

CONSULTING

For the second time in my career, I was working for a Consulting Engineer. I was a consultant. I had to fill in timesheets. I accepted the position of Senior Mechanical Engineer. This was far superior to Graduate Mechanical Engineer. It did seem strange to me to have become a 'senior' now that I had been out of university for six years. To be honest, I didn't feel very senior. But being an inexperienced senior was much better than being a very accomplished junior. The benefit of this title was that I was expected to be able to do my work without much supervision (which was good because I rarely handled close supervision well).

The major project that I undertook was called The Isolation and Lockouts Project. It was based at a local smelter. My job was to design and install isolation points on 50-year-old machines. This enabled the operators to 'lock out' the machines to do maintenance. It was a messy job since the old machines were not designed for modern safety standards. I was a Senior appointed as 'Lead Mechanical Engineer' (this is even better than Senior – Leads are picked from the seniors, particularly for their ability to lead a team). There was a Lead Electrical Engineer on the job and between us we had a team of one graduate engineer to manage. (Lexi was mechanical so she belonged more to me than to the electrical guy. He largely worked alone. Electrical engineers can be like that, it might be all the dividing by $\sqrt{2}$ that keeps them to themselves. At any rate, they aren't very often allowed to be in management except at electricity companies.)

It was a small but good team. The smelter had a 'three hours shut' (this is when the machines are turned off for maintenance) at 6am on Wednesday mornings. So we had a weekly opportunity to get the modifications done. During the week, we would do investigations, measurements, calculations, drawings and buy bits. On Wednesdays, Lexi would come into work early to supervise the work. We had both agreed that since I was Senior, I should be allowed to sleep in. (Well she hadn't argued when I proposed the regime.) It gave her some time to develop her skills in dealing with trades without me around imposing my views. She learned a lot, and I got to sleep in.

The project trundled on merrily. We were slowly but surely making the ancient plant almost human proof. I sort of became the project manager for this project. It happened because the Lead Electrical didn't have any desire to argue with our bosses or the client. (This is the job of a project manager; you always have two stakeholders to keep happy or as a minimum not upset irreparably. The client is the one who is paying you now. Your boss is the one who decides if you get a pay rise. The client wants as much as he can get for as little as possible. Your boss wants to charge as much as possible while delivering as little as possible - and make sure that any ongoing liability is minimised. Walking this line can be a delicate balance.)

My client was named Harry and my boss was also named Harry (this is a statistically unlikely situation. It wasn't that difficult to manage at the time, but does pose some difficulty in the telling). Harry (boss) rang me one day (he never rang any of the others, it was my exclusive privilege as PM). He told me of a problem that he had.

The situation was this: the three of us on my team were all employed on eight-hour-a-day contracts, but we were all working ten-hour days to fit in with normal working hours at the smelter. This means that we were being paid two hours overtime every day. It was glorious. We were also paid a contractual allowance for driving out to the smelter from our office: 71c per km for 10km each way added up to $14.20 per day tax free!

Harry (boss) said that Harry (client) wasn't happy to pay for our km allowance. (How tight can you be to deprive the poor consultants of almost 15 tax free dollars a day!) Harry (boss) said that we could no longer get paid the km allowance. I pointed out that it was in all of our employment contracts, and he couldn't modify the conditions of the contract unless we both mutually agreed. He got a bit angry at this time and said: "There's really only three ways this can go down: 1) the client pays the allowance, 2) we pay the allowance out of our own coffers, or 3) you don't get paid it."

"Either of the first two are fine with me," I replied, getting a little aggravated myself at this point. This caused him to back off a bit and try another tack.

"What about if we stopped paying you overtime... we could charge the client the overtime and use those monies to pay the travel allowance?"

I pointed out the obvious flaw in his thinking: "So you're suggesting removing a large amount that we are entitled to and are currently paid, in order to fund another small amount that we are entitled to and are currently paid. I haven't checked with the other guys but I don't think that anyone would accept that."

"Well, I don't know how we can resolve this." Harry (boss) made his veiled threat.

I had no patience left for this conversation and replied curtly: "Harry, you make your decisions as best you can. Then we'll make our future decisions based on your decision." The whole tone of this conversation stank. I went back inside the office to see Lexi and the electrical guy, who both told me that I'd said the right thing.

The next day Harry (boss) rang back and basically said that the km allowance was less than $15 each a day so he would just pay it (this was option 2 in the previous day's conversation). I thanked him for being reasonable, and we never spoke about it again.

One day when I was back in head office, Harry (boss) asked (told) me to work on another project urgently. He explained the project and how it was being managed (not managed at all) by the environmental group. There were a number of options on the table, none of which the client had endorsed or even commented on. But the pressing thing was that the client needed a project estimate so that they could make a capital request (a long and formal process of asking *the corporation's* accountants for permission to spend money in the future). "So, before you go home tonight," Harry (boss) continued, "can you knock up a project estimate that we can send to the client tomorrow?" I was unsure how to respond to this ridiculous request. I had no project background, and no idea what the project was. In fact, no-one had any idea what the project was, because none of the presented concepts had been endorsed.

"Harry, since we don't know what the scope of the project is, how do we think that we can provide a budget?" I gave a quizzical look, maybe because of the quizzical nature of my question. Or maybe because I couldn't understand why I was the only one who could see the obvious issue here.

"They don't need scope definition at the moment," Harry (boss) replied, "they can get that later. We just need to present a budget now, so that they can make a funding request."

But how on earth can you price a job when you don't know what the job is?! How is this going to end well?! He wants us to pluck a number out of thin air and put our name to it! Then in the future it will be discovered that the number is wrong and the client will blame us because we provided the number in the first place. Why are you people so stupid?! The emperor isn't wearing any clothes! My voice screamed within my head, while my outside voice calmly said: "I'm good. But I'm not that good."

Harry (boss) gave up at this point. He decided that it would take less of his time to guess the number himself rather than trying to get me to do the guessing for him.

Harry (client) gave us a dressing down for being late on the project schedule during a weekly meeting (with other parties present). We couldn't really dispute it. After the meeting, I was talking to Harry (client) one-on-one. He said that the project was going to get delayed soon, so it didn't really matter anyway. It got my back up. I pointed out that five minutes ago he was ridiculing me and my team in public and now in private he was saying that it didn't matter. He had a quick think on the spot and apologised. We got on really well from that moment on. Then he moved to Doha.

One day, *the corporation* rolled out a new overtime policy. Harry (boss) called a mechanical team meeting to roll out the new policy. In short, the policy was as follows:

o   If there was an opportunity to work overtime, do it;
o   Charge the client for any overtime worked (at the inflated overtime rate); and

o   You can only get paid personally for your overtime if your
     boss signs off on it.

He delivered this message as a good and faithful servant
(middle management) of *the corporation*. There were 14 of us in the
mechanical group. The old draftsmen raised their eyebrows but sat
silently. I felt that someone should voice the collective opinion of
the group. I said: "Harry, can I just ask, why would I work overtime
if I'm not going to get paid for it?"

"Well, it's the new policy," he offered. The old draftsmen
stayed silent, enjoying the show.

"Actually, I don't think it will affect me because I'm going to
resign later in the week." And I did.

I recognised a certain pattern in my career. Generally, I lasted
about three years in a job. The first year was spent learning the role.
New industry, new work, new stakeholders, new people. For the
second year, having learned how to ply the trade, I was content in
doing the job well. Discovering new ways to improve that others
had not. Somewhere in the 3rd year in a job, I realised that I couldn't
stand these people any longer, and I had to make a change. ('These
people' refers to the people who formed the management structure
above me. I can't remember a time when my colleagues or
subordinates caused the feelings of 'get out, get out now!') The cycle
was not always strictly three years; it did depend on the calibre of
management and how interesting the work was. Once I lasted almost
seven years.

This resignation occurred only nine months after starting –
not even one calendar year. If workers were frogs, jobs would be
like lily pads spanning the lake called Career. Some lily pads are large
and a frog could spend years traversing it before deciding to take the
leap onto another lily pad. Some lily pads were small and not
properly formed. Having jumped on, a frog might jump off as soon
as it could before the whole plant fell to the lake floor. Of course, if
all the frogs jumped off at exactly the same time, this might submerge
the lily pad. But the lily pads don't care; there are always more frogs.

PROJECT MANAGEMENT

I took a job with the water *corporation*. It started because I saw an advertisement for a Senior Major Projects Manager. It seemed to describe me pretty well – after all, I was now well embedded as a Senior, and although I had only managed one project, I had managed one project. The only sticking point was that the ad called for experience with the two Australian Standard (AS) construction contracts. I had never used them. So, I printed a copy of the two AS contracts and read them both cover to cover – well, I flipped through them cover to cover. I now considered myself an expert and ready for the job interview. When the inevitable question came up about the AS standards, I stated with confidence and truth that I was familiar with them both. To my delight, there were no further questions.

I got the job! It came with a car! It was a white Nissan X-trail with a massive number 80 stuck on the roof. This was not to separate the workforce from the senior managers, whose cars didn't have numbers on them. It was important in case I ever got stuck in the bush. It would allow the rescue helicopters to identify my vehicle from the sky. Apparently, we didn't care if the senior managers got bogged in the bush.

I worked with a guy called Tony Flower. But due to his very pretty surname, he was referred to as the Florist. (I don't think that he really liked it, but he didn't ever complain about it - unlike the Wookie). There was something familiar about him but I couldn't put my finger on it. After a month or so, I worked it out. From the hazy depths of my distant memory, I recalled that he had sort of introduced me to my darling wife. Years earlier, at university swimming club on a Sunday night, the babe, who is now my wife, asked me to check in the change rooms to ask the Florist if he was going running on Wednesday morning. I accepted the challenge. Entering the room, I asked the three towel-drying bottoms in front of me if one was called Tony. One turned. I asked him the crucial question.

"Yeah, sure. You should come too," he stated.

"Nah, not for me." I stood firm. Running wasn't my thing. And I certainly wasn't fit enough to do what these guys were proposing.

"Yeah, go on," he pressed.

"Ok." Wow, what an about face!

As I quickly left the changeroom since I was already dried and clothed, I wondered at how easily I had been turned. Sure, it had more to do with my now darling babe than the Florist. But still, I had been brought up on rugby, where you stand your ground no matter what! It turned out to be a good decision. On Wednesday morning, we went for a 14 km run. Yes, you read that correctly. 14 kilometres. 7 out and 7 back. I thought I was going to die on multiple occasions, but I survived. And after many more kilometres pounding the pavement, I got my girl. After that first introduction-run, the Florist never ran with us again. The next time I saw him was my first day at the Water *Corporation*.

The Florist managed multiple construction contracts and was really good at it. He knew all the contractors in our industry. He knew what they were each good at. He knew the people who managed them. And he knew the people who did the work too. He understood the power of relationships. One day our boss, Noel Jenkins (or just Jenkins, behind his back), told the Florist that he was lucky because none of his projects had failed. (Jenkins was a prickly kind of guy. By prickly I mean unlikeable. He was probably disliked by everyone who had met him, but more so by those who worked with him. Of this group, the contractors working for him disliked him the most. Let's just say that if Jenkins had ever been found dead in a dark alley, beaten to death with the bloody stub of a broken beer bottle, the police would have been interviewing all his contractors as suspects.) "It's not luck," was all the Florist replied.

The Florist was like the eye of the cyclone – wherever he went, calm went with him. In the office, and in the field. People were at ease with him, and he never seemed to have any arguments.

Jenkins, on the other hand, was a strange dude. Conflict was his travel companion. He was a micro-manager who didn't trust his employees. In return, his employees didn't like him. He would often come around and give us directions on how to do our jobs. He would start with the Florist, who would reply: "Ok." Jenkins took this to mean 'I have understood your direction and will comply with it fully forthwith! Sir!' It was clear to everyone else that the Florist's 'ok' was more along the lines of 'I agree that you said something.' It was weird – Jenkins thought he was in control, and without a skerrick of conflict, the Florist would not comply.

I, on the other hand, would argue with Jenkins for hours, explaining why he was wrong. He, for his part, explained at length why I was wrong. While Jenkins and I argued, the Florist was busy silently building stuff – peacefully. Being in conflict with your boss is exhausting. Eventually, I became resigned to the fact that I was really in a win-win situation. If my projects were successful, then that was good. And if my projects failed, then that would reflect badly on not just me but on Jenkins as well. That would also be sort of good. Making fun of your boss is a poor substitute for having a good boss. But sometimes it's all you have.

My first project was to build a water supply pump station to a design that had already been completed by Jenkins. Construction jobs begin by digging a hole to fill with concrete called foundations. The buildings sit on these foundations in order to not fall down. The design showed that there was mudstone nominally two metres down. (Mudstone is obviously a type of rock. I deduce that from its name. I don't know anything more about it. Maybe Carston knows what mudstone is after spending three years with Professor Milenko). The mudstone was discovered by a geotech drilling campaign during the design phase (managed by Jenkins). When the holes were drilled, mudstone was found about 2 metres deep. The foundation design was to dig down to mudstone and fill the hole with concrete sitting on the mudstone. Simple. The only problem was that there wasn't actually any mudstone or any stone at all. I took a call from my favourite site manager, the Wookie. (The Wookie had moved to a different construction company and was now building my pumpstation. It was like putting the band back together!). He said that they had excavated three metres down and there was no rock. I got onto the phone to Jenkins to explain the situation. He said that we needed to decide on a course of action that would involve a redesign of the foundations. Maybe 'bored piers' but that would take some weeks or months to design. That sounded bad. A re-design when you've already mobilised the construction crew onto the job is expensive.

The Wookie arranged a big drill to find out how deep the rock was. A day or two later we finally found it almost seven meters deep. We then dug a deep trench to expose the top of the rock and installed steel shoring inside the trench walls to stop the dirt collapsing into the trench. We found the geotechnical engineer who wrote the original geotech report to come to site and share in the investigation. We got into an excavator bucket and were lowered into the brand-new trench. It was like a very slow moving roller-coaster. (You're not allowed to do things like this anymore. Even though its safe, it could be deemed as unsafe. And being deemed unsafe is almost as bad as actually being unsafe.) We got out of the bucket and I asked the pivotal question: "Are we standing on mudstone?"

"I don't know." He replied.

"What do you mean you don't know. Aren't you a geotechnical engineer?"

"I don't know what this stone is unless I have it tested in a lab." That sounded like it would take a while. I thought that if I asked a better question, I might get a better answer.

"Is the rock we're standing on consistent with what you'd expect mudstone to be like?"

"Yes." Good enough for me. And not soon enough! As we stepped back into the excavator bucket, the muddy walls started to collapse below the steel shoring plates. Mud oozed about six inches thick across the mudstone floor. My boots were saturated and my feet were muddy and cold. My mind was taken off my discomfort when the geotech engineer told me that when they did the investigation back when the design was done they drilled the holes about fifty metres away and asked why we moved the design to this location. I found this very interesting. Why did we (Jenkins) move the location to this location?

I discussed the options with the Wookie and got a quote to just excavate the entire seven metres across the whole building area and fill it all up with concrete, basically applying the original design to the actual depth of mudstone. To my delight I had just enough money in my project contingency budget to afford this without having to ask *the corporation* for more money. I signed it up as a contract variation and the Wookie immediately got onto fixing the problem. All in all, we were delayed only two weeks. I thought it was a glorious success.

Jenkins was furious. I'm not sure if he didn't like that fact that I'd made a decision without him, or maybe he had made a promise to someone more important within *the corporation*. He never explained why he had the geotech investigation done in one place and designed the pump station in another location (that in retrospect looked very much like an old river bed).

◊◊◊

The geeks who fixed our computers were brilliant at *the corporation*. They had set up (networked) Castle Wolfenstein (a first person shooter computer game). Many a lunchtime was spent logging in and shooting at each other's avatars. There was a female graduate engineer whose avatar name was Princess Sparkle. She was quite a good gamer and had chosen this name deliberately because it was extra humiliating when the game said things like "You have been shot by Lieutenant Princess Sparkle," or "Rampage! Captain Princess Sparkle has made ten kills in a row!"

Once we did an all-nighter Wolfenstein session. Well, we planned to do an all-nighter. About 20 of us knocked off work early at about 4pm. This meant shutting down our emails and loading Wolfenstein – all achieved without even changing seats. At 6pm we had a momentary break for pizza, before getting back into the action. Princess Sparkle hunted me for another five hours or so until about 11pm, when we decided that we weren't all that young anymore and our own beds were probably the most comfortable places to be. We went home.

This workplace gaming habit gave my young kids a false impression of what the workplace was really like. It hit home just how good it was when my 4-year-old son said: "I'm sick, Dad, can I come to your work and play the shooter game?"

One day, we came in early at 4am to watch the Soccer World Cup final. I can't remember who won, and I certainly can't remember who lost. I do remember curling up under my desk mid-morning for a short kip. The Florist and I shared an office by this stage, and he too made a makeshift bed beneath his desk. I hoped that any visitors wouldn't hear any snoring if they came to the door. Visitors usually assumed that we were 'on site' if they couldn't find us. We were project managers after all.

One time, the Florist was out of the office for three days without telling anyone. When people came looking for him, I suggested that he was probably on site and to try his mobile. After two days, I rang his mobile myself. When he answered, I asked the million-dollar question: "Where are you?"

"I'm on site," he replied.

"You don't have to lie to me," I said, knowing that he couldn't possibly be on any of our sites.

"I'm in Townsville but I'll be back tomorrow. If anyone wants me, tell them to call my mobile and I'll get back to them." Oh, he was good! He could run his projects from another part of the country and no-one even noticed.

Suppliers who sell goods to engineering companies have a habit of giving their key contacts periodic gifts. These are typically small in nature like a bottle of wine at Christmas or a seat at a sports game. The idea is to say thanks for your business. Apparently, they can be very big. After all, it's not uncommon to hear of a politician who made an 'error of judgement' when accepting an all-expenses paid, month-long family vacation in the Swiss Alps as a gift from a land developer.

Companies have rules about accepting gifts from suppliers. These rules typically take one of the following forms:

o   Never accept anything ever, and report it if you are offered anything;

o   You can only accept gifts under a certain (small) dollar value and a maximum of three gifts per supplier in a calendar year;

o   Just declare anything that you're given.

Gerry bought valves. Before Christmas, one of the major valve suppliers phoned him to offer Gerry a bottle of scotch in thanks for his business over the last year. Gerry said: "Thank you. I'll let you know which one." I'm pretty sure this is not how the salesman had intended for the conversation to go down. Gerry would then do some research and find a particularly obscure scotch (not necessarily expensive, just hard to come by) and hence send the poor salesman on a wild goose chase trying to find a bottle of Glen Something-or-other from the Isle of Sky. I was witness to one Christmas Eve when the salesman got to Gerry's office late and was very apologetic.

"I've been to every bottle shop in the state and the only scotch I could find that resembled your request is this Glen Thingamy. It's distilled on the Isle of Man so not quite the same but I think it's similar." I was amazed – to Gerry, the aim of the game was to ask for more when offered a gift (I'm pretty sure this didn't comply with the company policy, even though I never read it). His goal was to find the point when the salesman would say no. Gerry kept the bottle of scotch for ten months before giving it to his father for his birthday in November the following year. He didn't even like the stuff. Every December, Gerry and the salesman did the scotch dance.

We bought a lot of pipes. Water infrastructure is largely combining pipes (in specific configurations). Thin pipes, thick pipes, steel pipes, plastic pipes and even iron pipes with plastic coatings. Even the tanks we built are like short and very fat pipes sticking upright instead of lying flat. There was one main pipe supplier who we bought most of our pipes from. The sales representative from the pipe supplier, would come around every couple of months to get an update of potential future orders and to get to know us better. He was based in Melbourne, and we all got to know him pretty well. One day, when a colleague and I were booked into a conference in Melbourne, I rang the Pipe Rep and asked outright: "What have you got on Friday, five weeks from now?"

"I can't do anything with you guys, since I'm taking people to the footy for the Carlton vs Collingwood game at the MCG."

"Great, put me down for two!"

Pause. "I can't." It was the only time I can remember the Pipe Rep losing his typical salesman's charm. "I'm not allowed to spend more than $80 on anyone who works for any government body." I had done it. I had achieved Gerry's aim and found the point when the salesman said no.

"Well, we're not allowed to accept. So, you don't tell your people and I won't tell mine. See you at the game." Again, I was not exactly sure about the details of our company policy, but I may not have been fully compliant.

Carlton losing was the only downer on that night at the footy. When we arrived at the Corporate Box, the Pipe Rep asked what we thought of their box. It was magnificent. Standing in the open window above the thronging masses I declared: "It makes me feel like I'm the king. I could lean out and spit on my subjects from here!"

"Oh, please don't!"

◊◊◊

I was bored building infrastructure (and by building, I really mean employing other people to supervise the people who actually do the building). It just wasn't exciting me anymore. I realise now that this is because engineering is fundamentally boring. What to do? In times like these, you go to what you know, which in my case was effortlessly passing university courses. I went back to my old Uni and enquired about doing a PhD. You have to understand that this was at a time in my life when I still thought PhDs were smart (and pilots were cool). To fully consider the merits of trading work for study, the following four questions need to be answered:

1.Am I eligible to do the study?

2.What course/field can I study?

3.How much (or little) does it pay?

4.Do the lifestyle benefits outweigh the lack of money?

I discovered the following four answers:

<u>Answer 1</u>: Since I had attained Honours with second class uppers, I was eligible to go straight to a PhD without having to do a Masters degree first. So far so good.

<u>Answer 2</u>: They had a piece of work to refine one of the angles in the design of a ship's bow (the bit that hits the water) that was sponsored by a local company that built boats (the fact that it was sponsored is important in Answer 3). A recently completed PhD had established that there were some benefits in increasing the angle. This new PhD was to determine what the best angle would be. It would involve a lot of hydrodynamic lab work (hydro-dynamic lab work is awful – I remembered it from 3rd and 4th year the last time I was at university; really hard analytical maths, with lots of stand-in letters instead of actual numbers [like π but much more complicated]). Hmmm. This sounded even more boring than building stuff. But I pushed on to address the next questions.

<u>Answer 3</u>: Not much. But because this was an industry-sponsored PhD it paid about twice what a non-sponsored PhD would. This was much simpler maths. I could work out that we could just manage on a PhD income, but it would be very tight.

<u>Answer 4</u>: The case *for*: you don't have to really go to work. You can just, sort of in your own time, go into the lab, design experiments, hopefully work out what the letter-numbers really mean and record lots of measurements. After two years you would come to staggeringly good conclusions and write them up into a best-seller that presented these results in a way that was as elegant as it was accurate. Every boat designer in the world would buy a copy to keep

on the shelf with the really useful design manuals. The case *against*: you had to go back to university, get paid very little, and do research.

Conclusion: What was I thinking – I can't research! I can't even concentrate for more than about 12 minutes at a time. How could I possibly extend this attention span to last two years? And although competent at maths, I don't like doing it. I like talking to people, not doing calculations.

So, I came to my senses and went back to work. Occasionally, when I meet people who have PhDs, I say that I was offered one once. Sometimes I add that I didn't have what it takes. I am really a generalist. My speciality is not having a specialty, but knowing enough about everything to know when to engage the specialists.

(There was a time many years later, when working on a project review team with other industry "experts" in their fields. It occurred to me that four of the team of six had PhDs. Four PhDs, me and an estimator. I brought this to the attention of the team. (Dr) Dennis (the oldest PhD there) laughed and said that he thought that the only thing that his PhD really did for him was reducing his salary for the last 30 years. I don't think he fully valued having Doctor on his airline boarding pass. Mind you, Captain would have been even better.)

As boring as construction was, it had a lot of things going for it – like I got paid every fortnight, and I had a company car that I could drive around in. It seemed that boredom was the price of comfort. Little did I know that the boredom was soon to be banished.

After a couple of years of being a Senior Major Projects Manager, the whole water industry got restructured. This involved fourteen departments from different local councils being combined with us into another entirely new *corporation*. It was chaos. I loved it.

During the restructure, we were relieved of having to report to Jenkins. That was nice. The Florist and I were left to run the whole capital works (construction) program for the new organisation. The transition was messy; each of the organisations that had been merged had made a flurry of commitments preceding the merger (to ensure that their projects couldn't get canned by the new *corporation*). These commitments weren't very well documented. In fact, many were not documented at all. We generally found out about a commitment when a contractor rang us up asking to get paid for some work that they were doing. We would ask to see evidence that they were contracted to us, and they would produce their contract (then we'd photocopy it). In the beginning, we had a 'surprise!' every day. After a month or so, it calmed down to about one a week. The best (or worst) surprise that we found was a purchase order to a contractor that read like this:

Item	Description	Amount ($)
1	Dam Construction	TBA

This was inconceivably sloppy. It took months to back out of that one and put in place a proper construction contract that included sophisticated things like a design, a price and a schedule.

At this time, I got to know our lawyers. (Incidentally, have you ever pondered the difference between a Barrister and a Barista? An English Barrister has joined the Bar in order to practice law in the court room. An Italian Barista stands at the Bar to make Coffee. The two cultures think very differently about the nature of "the Bar.")

Lawyers, or "counsel" as they are known, come in two categories: internal and external. Internal counsel are just like other employees, but instead of having an engineering degree they have a law degree (and they wear a suit and can't build stuff). External counsel, on the other hand, work for a law firm that is engaged by *the corporation*. They are sophisticated, high-flying, influential and manipulatable (they also wear suits and can't build stuff). I was once doing a complex series of land acquisitions so we could build a pipeline. At one of my meetings with our external counsel (Brenda and her assistant Julio), I was very clear in my request:

"Brenda, I want legal advice that says that I can do xxx." I can't remember the details of my request.

"Jack, we can't give you that advice…. because it's not legal. It's actually illegal."

"Oh, really, do we have to change our external lawyers then?" I replied with a smile.

"Don't start that again, Jack," Julio chimed in.

"Look, Jack," said Brenda, "there is another way you can go about this which will get you to the same result, and it's legal. I can give you that advice in writing."

"Excellent, thanks for your time. If you guys work on that advice now and email it to me, I should have it in my email by the time I get back to my office," I replied. They both rolled their eyes and Brenda pushed the file over to Julio's side of the desk.

◊◊◊

We were all forcibly trained in the Safety Circle™. I can't remember which silly consulting (or marketing) company came up with this name, but they convinced *the corporation* to engage them to spend 6 hours with all of the employees making sure that we understood their marketing gumf.

I came to work one morning to find out that I was expected in Meeting Room B and I was already half an hour late. When I walked in there were already about 20 of the operations guys (trades - who do real work, wear yellow shirts and start work earlier) sitting around the tables being trained in the Safety Circle™. I didn't get on with the trainer from the silly consulting company right from the start. He asked me what time it was. I replied that it was starting work time. I don't consider that I had to justify my movements to an external consulting trainer. It was obvious that the displeasure was mutual. After sitting through five and a half hours of this guy (who neither did work or supervised those who do work) teaching us how to work safely, it was time to make a commitment! He passed out blank certificates and invited each of us to make a personal commitment to "working within the Safety Circle™." The way that you did this was to write your name on the top and sign the bottom. The certificates were then collected as evidence of our 'commitment.' The choice was easy, sign the form and then you'd be allowed to leave the training room. I ummed and ahhed a bit, but

I couldn't do it. The guy was jerk, why am I even here, I should have just ignored it this morning (my day and life would have been better). The trades were busy signing and handing in their certificates.

"I'm not going to sign it." There was silence. The guy was obviously not prepared for someone to say that. There was a little more silence before he responded.

"You don't have to sign it, Jack, if you don't want to work here." I thought that he had significantly overstepped his authority at this point. I dug my heels in further.

"So if I don't sign my name to your proprietary name: Safety-Circle-trade-mark-registered, then this organisation is going to sack me? I'm pretty sure that that isn't allowed under Australian employment law. And just so you know, I've never had an injury on any of the projects that I've managed. And I don't intend to have any in the future either. I'm committed to working safely as I always have. Are you seriously saying that signing up to your Safety-Circle-trade-mark-registered is more important than my performance?..."

In the end we compromised and I crossed out 'I commit to working within the Safety Circle™' and handwrote 'I commit to work safely.' I initialled the change, and demanded that the jerk also initialled it before I signed the bottom of the certificate. We were both much relieved when I left Meeting Room B that day.

At this time, I took on a mentoring role which involved me mentoring other project managers. One of the biggest things that I worked on with them was the concept of 'being comfortable with being uncomfortable.' During any project, there are always loose ends and issues that need resolution. The only time when all issues are resolved is after the project is completed. A good project manager knows which issues require attention at what time, and indeed which ones to delay responding to or even ignore. There is always some discomfort during projects. Hence, the need for a PM to be comfortable with being uncomfortable.

Wow, I realised that I had inadvertently got good at this. I wasn't pretending any more – I actually knew how to run construction projects. People came to me for advice.

After all the merger-induced excitement had settled down, things were largely under control. If I worked hard, I could get my week's work completed by about 11am on Tuesday. The trick was to not work hard, so that you could spread it out over about four days. Sometimes the Florist and I would go to the movies in the afternoon. People in the office would assume that we had gone to a work site. I guess people at the box office assumed that we were bums without jobs. As nice as having a job that is not challenging sounds, after a while I found the lack of purpose soul-destroying.

The perennial question crossed my mind again: what else could I do that wasn't construction project management? I had an idea: I could become a doctor! It sounds crazy I know, mainly because it was a truly crazy idea. But, not one to let sanity stand in the way, I went back to the university to investigate the possibility of doing medicine and becoming a doctor. In a lot of ways this was simpler than my last foray back to academia (the event now enshrined in my family's collective memory as 'the PhD debacle'). The rules in the medical school were a little different to the PhD application. All I needed to be accepted into medicine as a mature-aged-student was a distinction average in my first degree. Sounds simple, right? Well, having found my academic transcript, I asked the university if I had indeed attained a distinction average. Let me just say that if I hadn't got that terminating pass in 4th year, it would have been clean cut. Looking at my transcript, I realised that my marks had progressively decreased the longer I stayed at university. I had a high distinction average for first year, a distinction average for second and third years and something in between a credit and a pass average for fourth year. (Following this pattern, if it was a six-year degree I would have failed it!) In the end, the clerk took a copy and sent it away to someone more academic to make an unequivocal judgement about what level my university average was. This judgement took two weeks. I imagined that they had to wait for a full sitting of the Vice-Chancellor, Deans and Heads of Department to debate my transcript (this probably would have occurred in a wood panelled room with high backed leather chairs and a fireplace). In the end a judgement was reached: I was deemed to have attained a distinction average. Yay, I could become a doctor (and pay them more fees).

Having fought the fight to establish that I was eligible for admission to Medicine, I now turned my mind to the other salient factors. It would take at least seven years to get fully qualified (wow that is a long time!). I would have to study again. I would have to study stuff that I had never done before, like biology. For the first four years I wouldn't get paid… What was I thinking!? This was 'the PhD debacle' renamed – 'the Medicine debacle!' So, once again, I went back to work, wondering if there were any other prestigious university courses into which I could get accepted and then decline.

(Some years later I stumbled across a practice Mensa IQ test on the internet. I did it while sitting on the toilet. To my delight it recommended that I enter myself to do the real test. There was no way I was going to do a real Mensa test. If I didn't score well the excellent result from this practice test would be superseded and wasted! Instead, I just told people that I was a borderline genius based on the online practice test. My brother pointed out that a borderline-genius can also be just called borderline-normal.)

◊◊◊

Some months after the merger, I received an email from a new girl in HR, asking me to reply to the email with my name and position title. Seemed a little strange. Upon further inspection I saw that it was an all-staff-email asking everyone to reply with their name and position title. Although most people replied as instructed, I saw an opportunity. It was obvious that the new HR didn't have records for all the employees (if they did, they would know what everyone's title was). I promoted myself to Project Director that day.

There was one other legendary all-staff email. It was from a lady called Harriet in the document control area. Harriet sent an email to all staff, asking if anyone had a copy of the book *Richard the Lionheart* for her son's 9th grade literature class. I thought long and hard. Will I? Won't I? Will I? Won't I? And then I did.

Reply-All:

Dear Harriet,

Unfortunately, I do not have a copy of this book, good luck in your search.

Cheers,

Jack Martino BEng (Hons), CPEng, NPER
Project Director

Send.

Then I waited. Giggling inside. I thought that I should probably be punished for this, at least a stern talking to. The Florist giggled out loud.

After three minutes everyone got another email:

Hi Harriet,

I don't have the book mentioned, but I can recommend another named "Lionheart" – it's more historically accurate. I have a copy and can lend it to you.

Regards,

David - Asset Management.

Then another:

Harriet,

Can I recommend the Lord of the Rings, it's a great story and it's really three books. A lot of bang for your buck.

Kind Regards,
Gerry - Engineering.

Later that day, all-staff email functionality was removed from everyone below the most senior managers.

Although it was delightful to have been relieved of having to talk to Jenkins ever again, we now had a new CEO who was dubbed '*Write-Me-A-Memo* Mark' (his name was Mark). He had never been a CEO before and he wasn't very good at it. Whenever we needed his approval or direction for anything, he would reply with "write me a memo." Now, some things are memo-worthy, but some things are not. For example, requesting signatures to execute a multi-million-dollar contract is memo-worthy. Requesting an update on the new seating plan is not.

I took the opportunity when *Write-Me-A-Memo* was on holidays to award a variation to a contractor that was three times the size of the original contract. (Although not illegal, this is frowned upon since variations are thought to be small and minor additions to the underlying contract). It was a good move since the contractor was performing very well and a timely award was needed to hold project schedule. It was signed off by *Write-Me-A-Memo's* delegate. I knew it was coming, but it took three months for *Write-Me-A-Memo* to find the award and summons me to his office. He ranted for a while, then culminated with: "...... and who approved this variation?"

"Well," I replied, "it's your authority level, so it could only have been approved by you or someone that you delegated." He looked at me, infuriated. I chose to leave his office at that point and we only spoke again once.

Write-Me-A-Memo (together with the executive team) decided that there was to be a new salary policy. In a nutshell, the new policy was to pay everyone but themselves at the 25th percentile of wages (this means that if there were 100 people in the state doing the same job, the one who worked for *the corporation* would be paid less than 75 of them). I guess that this policy was good if you wanted to attract and retain the lower end of the market. Since I was a Project

Director, I didn't think that the new policy was designed with me in mind.

The Florist and I had a chance to speak to *Write-Me-A-Memo* about it in our office one day. I asked simply: "Do you believe that the 25th percentile policy is going to benefit *the corporation* long term?"

"I think it's fair and I think it's generous." What a pompous goat.

"Well," I chose my words carefully, "it doesn't suit me, so I'll go and work elsewhere, I guess".

Write-Me-A-Memo responded in a strange way: "Here, you have security of employment; we'll always have a job for our employees regardless of whether the times are good or not. If you want to go and play in the risk game, then go for it. But you might find yourself out of work."

"I'm not scared, Mark."

It was the last time we spoke. About a month later, I left for the 'Wild West' of engineering: Mining. Little was I to know that I was about to join 'the risk game' as *Write-Me-A-Memo* had so aptly put it. (As it turns out, mining is a wonderful industry. Made of big but simple projects normally delivered by big and simple people. It doesn't take much to stand out. In most cases borderline competence is all it takes.)

MINING

I took a phone call on a Monday morning from Harry (who used to be my client at the smelter). It was 8am in the morning. Harry was upbeat and he said: "I've got a job for you. When can you start?"

"Where in the world are you at the moment, Harry?" The last I had heard of Harry was three years earlier when he moved to Doha in Qatar.

"I'm back in Australia, working for this mining company. I'm PD for this multi-billion-dollar project. I've got a spot for you. When can you start? Do you have any leave that you can use in your notice period so you can start sooner?" (PD stands for Project Director; it's basically a Project Manager who manages Project Managers.)

"Well, I'm interested, I'll talk to Bella (my wife) and get back to you." I could feel that this was about to start moving really fast.

"Ok, talk to Bella, then come over for an interview (and by interview, I mean to sign a contract) and do it this week because I have to go to China next week!" It was arranged. We were on a plane that Friday for a job interview. I signed a contract the following Tuesday and I started work two and a half weeks later. What a whirlwind! That is how you recruit. No ifs, no buts, just get excited and get started before any of the excitement wears off.

(I had leaving drinks with my colleagues from the water *corporation* before leaving. It was on a Wednesday afternoon at a pub on the waterfront. Wednesday is a strange day for leaving drinks, but it was selected because the pub in question had a "Uni night" on Wednesdays, where 'poor' students (without jobs) could load up with cheap beer. None of us were students (and we all had jobs), but on Uni night everyone was treated as a student. At 4pm a stein of beer cost $4. A Stein is a big German beer glass of 1.5 litres. At 5pm the price increased to $5. At 6pm the price skyrocketed to $6. I can't remember what happened to the pricing after that, but I have my suspicions based on a certain recognisable pattern. I got there at 4pm on the dot and ordered a beer. My buddies filed in over the next couple of hours. It is customary to buy someone a beer on their birthday, or other special occasion like if he's resigned. I didn't really think about this when I picked the pub. Let's just say that if you buy someone a 10-ounce beer (285mL) he can probably drink it without any severe side effects. However, when you buy someone a stein (1.5L) he may or may not be able to finish it. Or if, say, a couple of friends each bought him a beer, he might have a stein in his hand and another 3 litres of beer on the bar next to him. That's pretty much what happened to me. I don't know how far through my 4.5 litres of beer I got through, but I started feeling a little light headed around 6pm. So went outside for some fresh air. It seemed natural at this point to lie down on the wharf concrete. The cold concrete seemed much more comfortable than standing up. Sometime later I was woken to a tap, tap, tapping sound close to my head. I opened my eyes to see a seagull eating from a puddle of vomit next to me. Strange. The next thing I remember is waking up at the end of a table in a restaurant. Around the table were most of my closest friends and colleagues. They all had half eaten meals in front of them and they seemed to have been having a jolly old time. Intriguingly there was a half cup of cold espresso coffee in front of me. Feeling quite peckish, I got up and picked up the half steak that was on the Florist's plate. We had dessert after that, but I passed on the offer for port.)

Working for Harry was great. We got on really well and have remained friends ever since. I was discovering a whole new industry, a new state, and big project culture. Multi-billion-dollar projects are like mid-sized companies in their own right, with their own dedicated staff. This one occupied a whole floor of a building in town, as well as 2000 people on site (actually doing the work). I was brought on as Principal Project Engineer (interesting title – principal is better than senior or lead) and I was working for a Project Manager (or PM). Within two months, I was doing half his work and was promoted to PM in my own right. (This is the only time I have <u>ever</u> been promoted within a *corporation*. Every other time, I've had to leave in order to get a better job.)

We were building all sorts of things: roads, camps, electricity supply and generation, and water infrastructure, as well as getting really big diggers and trucks to make enormous holes in the ground.

This was mega-project world – it was mining. Low-tech, high-pay, fast-paced and unsophisticated. It was all going swimmingly well until Harry got sacked one stormy night and everything changed.

There was a meeting one evening in *the corporation* offices. Harry, his boss, and his subordinate, the Deputy Project Director (DPD) all went into a meeting room. Three went in and only two came out. Harry never spoke about what happened in that meeting. I imagine that the DPD said something like: "This project ain't big enough for the two of us, hombre. Step aside cos there's a new sheriff in town" as he spun the barrel of his 6-shot revolver before firing six times. Harry, lying in an ever-growing pool of his own blood, stammered: 'but…I…trus…ted…you" before expiring. The big boss silently watched the whole scene then turned to the DPD saying: "You did good, Son."

Suffice to say, Harry left the business that night, never to return. (Harry went to work in Africa doing more mega-mining projects. A couple of years later, he decided that he had earned enough money for his lifetime and retired to a farm in country NSW at the ripe old age of 42. So, being "assassinated" by the DPD turned out for the best in the long run. But that's another story.) The DPD took over the project. He was a fool. The project was doomed. I started looking for another job. This was the shortest job I had ever taken – I had lasted nine months at *the corporation*. It was a brilliant ride and got me established as a mining professional during the mining boom in the midst of the skills shortage. It turned out to be the perfect storm for me.

Since I wanted another job, I got an agent. In the heady days of the 'mining boom,' everyone including engineers had an agent (not just actors). I asked people I trusted and found an agent that I was happy to represent me. I was a bit cautious because I didn't want a used-car-salesman type of agent. The one I got was good. She was a genuine person with good contacts. We had a strategy meeting in a café; she improved my resume and she put me forward to a number of companies who were advertising. She would be paid by the recruiting *corporation* to the tune of 15% of my first year's salary if I was successful (so I didn't feel bad about the 'free' resume advice).

It is a strange phenomenon that many senior managers feel comfortable with a bully running construction projects. I have come across many in my career. People who seem to think that issues will disappear if you yell at someone. It seems to give senior management comfort when they see the angst because they interpret it as everyone trying as hard as they can – when the reality is the opposite. On the one hand, if there is no pressure to deliver, then things get very lackadaisical. No drive means that people have no direction and this leads to complacency and poor delivery. On the other hand, if there is too much pressure and everyone is stressed because they cannot achieve what is being demanded, then people start taking risks (which rarely come off). I haven't met anyone who is at their best when under constant, overbearing pressure. There is a sweet spot where everyone is under some pressure but not stressed. And if you can get the team into this sweet spot then the team can achieve amazing things. But to senior management, this probably looks like the people aren't trying hard enough – observable by the absence of stress.

As my time at *the corporation* came to a close, we had a weekly status meeting first thing on Monday mornings. There were 12 PMs present who were each grilled by the DPD as to the status of their projects. He demanded that everyone deliver their packages early! There were all sorts of excuses tabled for delays to his demanded dates: most plausible, some less plausible. Everyone got yelled at in turn. After two hours of public castigation, we all left the room suitably chastised. (Indeed, sufficient chastisement to last the week until we re-assembled for the next round of this perverse sport the following Monday.) Senior management could rest easy, sure in the knowledge that everyone was doing everything that was humanly possible to deliver early. In fact, the meeting had zero effect in changing the delivery plans of anyone in the team. It was just one of those things that one had to endure if one held the lofty position of PM.

Seven of the project's twelve project managers were seconded from an engineering consultancy. One day, the DPD decided that he didn't like the consultancy any more, and he sacked all seven in a day. He then set about recruiting seven PMs into the vacant roles. It was a massive knee-jerk reaction, and hard to say how it was in the best interests of the delivery of the project. One of the old and jaded contract engineers on the job described it best: "I've never seen this before! I mean, I have seen management teams replaced, but usually they bring in the new guys first and get them familiar with the project. Then they sack the old guys. This is new!"

One of the new PMs who joined the team was a controls engineer. (Controls engineers are sort of electrical guys who aren't allowed to work on real electrical stuff, so they connect computers and devices together in order to make mechanical stuff operate.) After about a week at *the corporation*, he took me aside and pointed to the full-wall project schedule that was stuck up in the project office, saying:

"Why is this schedule up on the wall, considering that not a single project manager in this office believes it's possible and has neither the ability or intention of delivering to it?"

"Yeah, it's pretty messed up. On a brighter note, I'm resigning later today!"

"No, seriously…"

"Yeah, seriously." At this point he gave up trying to get any meaningful answers from me.

A year later, I heard on the grapevine that the DPD had himself been sacked by management. Maybe he hadn't yelled at his quota of people since I left. Ah well, I suppose if you live by the sword you die by the sword. Eventually, this project was finally finished about a billion dollars over budget and about a year late. I wondered if the schedule was still up on the wall.

THE GREATEST COMPANY ON EARTH

I was recruited by a multi-national mining giant with operations in many countries. It had been digging up rocks to sell for about a hundred of years. It was the greatest company on earth. This wasn't the slogan or anything, but it was the vibe.

I was interviewed by a General Manager (or GM for short) named Joe Mutese (later on, like everyone else, I called him Mutsy. He didn't seem to mind). I was confident that the interview was successful when it deteriorated into us swapping stories of driving in southern Italy (I don't think that the HR lady had driven in southern Italy since she didn't share any stories).

It took a mere four months from that interview for me to start work. This is lightning quick for this *corporation*. Most of my colleagues waited at least six months before being accepted into the fold. But there was a mining boom going on, and I suppose that caused them to hurry things along for me. The boom was fantastic. It meant that young people like me could become really important really quickly without necessarily knowing much.

I was employed as a Project Manager. This is very important because *the corporation* was like the British Army in that it was very hierarchical. I didn't know how important it was to be employed as a manager or, as we said, at *manager level*. The first time I went to site was with a Study Manager (*the corporation* was very careful to make sure that the minions could tell if you were at *manager level* because you had *Manager* in your title). As manager-level officers of *the corporation*, we did what was expected of us in times like this and booked ourselves into the Lodge. In the key towns where *the corporation* operated, in addition to the cattle-type accommodation for workers, there were Lodges. The Lodge was a group of four apartments on the edge of the town, in the area where the manager houses are located (the manager houses are the nicest houses in the town). Compared to the usual mining camp accommodation, the Lodge was magnificent. The two-bedroom apartments were immaculate, spacious, and more importantly, maintained. It also had a freezer full of microwave meals, a shelf full of little baby bottles of whiskey, gin and rum, and a fridge full of dry, tonic and coke.

Up until this moment, I had always maintained a socialist leaning view of the workplace. I had never understood the way senior management would band themselves into the club of the privileged and treat themselves better than their employees. But as I enjoyed my scotch and dry nightcap, clearly now part of it, I began to think that maybe the club wasn't so bad after all.

Mutsy was good to work for. He was a great guy who had a human-centred approach to work. He cared for his people whenever he was allowed to. He just had a way of making you want to do a good job for him. You didn't want to let him down. And he owned his own winery. He was after all a Sicilian. Need I say more.

To start with, I didn't really have much work to do, so I made sure that I wasn't in the office being bored. I took a call from Mutsy's secretary one afternoon who asked me to come over to Mutsy's office because he had to see me that afternoon. She suggested a 2pm meeting. As it was 1:30, I thought that was a bit too soon for me to get to, so we agreed that I could be there by 3pm. Upon hanging up my mobile phone, in my kitchen at home, I got dressed back into my work clothes and drove back into the city. When asked, "Where were you?" I replied with "I was off site," which was a true and accurate description.

Mutsy had an urgent letter for me that he had to deliver to me that day. It was notification of my bonus. The bonus system here consisted of you getting a score between 0 and 200. In my contract I was eligible for a bonus of 20% of my base annual salary. A score of 100 would yield a 20% payment. A score of 0 would yield a payment of 0. And a score of 200 would double the contractual entitlement and deliver me a whopping extra 40% of my base pay. I presumed that since I only worked for *the corporation* for two months of the previous calendar year, I would not qualify for any bonus for that year. But I was pleasantly shocked to find that since I had not been at *the corporation* long enough to be judged thoroughly, I was awarded a score of 100 (and my bonus was pro-rated for the two of twelve months that I was there). What a generous company! Well, as my years progressed, I realised that the bonus system was actually about deferring salary. Although in theory the scores could range from 0 to 200, in fact everyone always scored between 90 and 120. So, my score of 100 for being unjudgeable was pretty much spot on, and I scored about this same number every year regardless of how well or poorly my projects had performed.

It is also interesting that the bonus was paid at the end of March for the assessment of your performance in the previous calendar year. So, by the time it was paid, you were almost at the end of the financial year (tax time). If you were thinking of leaving, you should really hang on till then. And then by the time you got your tax return back, you were more than half way through your next bonus assessment year. The 'system' was actually designed to stop you from thinking about leaving because there was 'bonus money' that you felt like you were owed. Deferred Salary. Clever.

My first project was with a Texan engineering company. They were pretty big in oil and gas but hadn't done many mining projects. There were so many projects on the go at *the corporation* that my project team couldn't fit in our offices. (The usual practice is for the consultant engineers to sit in the client-provided office space.) So, I relocated myself into their corporate offices with them. I was allocated an office with the project team outside my door. This is an unusual situation – basically I was totally unsupervised and unsupervisable. Since I was the client (who paid the bills) there wasn't anyone in their building who could tell me off. And as they say: power corrupts, absolute power corrupts absolutely.

The phone in my office had the name of the last guy who sat there. I walked outside my door to the project secretary and asked how to change the phone name. She told me that there was a form (there is always a form). She found the form and filled it out for me. She asked:

"What name do you want it to say?" Instead of saying Jack Martino, in a sudden burst of inspiration, I asked back:

"What do you think it should say?"

"I dunno," she replied, "how about *Big Daddy Jack*?"

And Big Daddy Jack it was. I was particularly chuffed when I could tell senior managers from the Texan company that if they wanted to find me: "Just look me up in your system under Big Daddy Jack."

I had been on the books for almost six months, quietly working out how to do my job when I was given to a new GM. It wasn't just me, four of us Project Managers (or PMs) were handed over to a new guy who had just been employed as a GM. This was a little unsettling since I liked working for Mutsy. Apparently Mutsy thought he had already told me that he was only acting as my boss. I don't remember that conversation. The first I knew was Mutsy's boss telling my contractor that Mutsy was being replaced by the new guy, Angus. It could have been worse: I could have found out after my contractor, or indeed my contractor could have told me that I had a new boss, or I could have read about it in the newspaper.

This was a bit of an issue for me since all I want to do is work for a good boss. I would go out of my way for a good boss. I'd work late for a good guy. I'd even work on weekends for a friend. Conversely, I really struggle to give an idiot the time of day, let alone my best.

I asked Mutsy about the new guy. I asked if I'd like him. Mutsy replied: "Why do I care?" He smiled then and said: "You'll be fine." Encouraging.

The new guy, Angus, was an aggressive, gambling alcoholic. In short, he was Scottish.

During my first meeting with Angus, he told me that he had to review my resumé, so I should send it through. Shock. What right did he have to review my resumé? I had already been selected in my position. I was terrified that he was going to take away my job.

It took me a week to tidy up my resumé, making sure that the embellishment was spot on: not too small; not too big. I sent it through. Silence. I took no news as good news and kept on going about my business albeit somewhat nervously.

I was a little anxious by the time I had my next meeting with Angus. We went through status and progress on my project with no mention of my resumé. Right at the end of the meeting, I plucked up the courage to ask: "Have you seen my resumé?"

"Yeah, I saw it the other day when it came in." He didn't seem to be taking this as seriously as I was.

"And…what did you think?" I stammered.

"You're a young guy with experience on one big project. Keep it up." I was relieved but still a little confused as to why I had been singled out to have my resumé reviewed. I worked it out a few weeks later when Angus sacked one of my colleagues. It became apparent that Angus had demanded a resumé review for each of his reports.

He went on to sack, remove, make redundant or fire 13 Project Managers and Project Engineers during the next calendar year. That's one a month plus one extra. And this was still the boom times. Angus showed his hand as being intolerant of poor performance. He would prefer to have no-one rather than a poor performer. In fact, on two occasions upon removing a PM, he appointed himself down into the position, and took the reins himself until he recruited a replacement. I had never seen a senior manager who was prepared to 'work down.' He was all about doing whatever it took to be successful. After our rocky start, we got on fantastically. We both thought it was lovely to work in an industry that had such low expectations that all you had to do was deliver on time to be a superhero.

After some time working for Angus, it became clear that he had another company on the side. It felt like he was cheating on *the corporation*. In one of our weekly meetings, I asked him about it. It was a small consultancy of about 10 guys that worked in Oil and Gas asset management (this is code for managing long lists of valves and their characteristics). His company was called Black Bull Consulting. The company logo was a big black bull with big white horns.

"Why did you name your company 'Black Bull?' Is it a play on words with your name, Angus, and the cattle breed?"

"I hadn't thought of that," he chuckled. "Bulls are thick skinned and charge a lot. Just like me." After all the first rule of consulting is: 'even if you're not part of the solution, there's good money to be made by prolonging the problem.'

"Do you have the standard 'exclusivity of employment' clause in your contract like the rest of us or did you negotiate something better?" The standard clause basically says that you can't work for anyone else while you're working for *the corporation*.

"My contract will be the same as yours, just with a bigger number in the salary section," he replied.

"Then how do you get away with managing another business on the side?"

"I don't manage another business," he replied slightly perplexingly.

"You own another business!"

"No, I don't. My wife does."

"And you reckon that you don't work in the business when you clearly do."

"I'm a director." Ok, now we were getting somewhere.

"Ok, so is that allowed? Did you have to declare it as an interest or something?"

"As far as I'm concerned, I've declared it." He said somewhat cryptically.

"So have you declared it or not?"

"I included in my *corporation* profile that I'm a member of the Australian Institute of Company Directors. And everyone knows that you can't be a 'member' unless you hold a position of director for a company. So as far as I'm concerned, I've declared it." Although it wouldn't stand up in a court of law, as long as his boss liked him, it was probably sufficient a defence if he was ever busted by HR.

Angus told me later that he had come to Australia to work in the oil and gas industry. He had come from Scotland via the Middle East (wow, two of the three global oil and gas hotspots). I asked him how long he worked in the Middle East. He said that he never worked in the Middle East: "I came via the Middle East."

"So, you imply that you worked in the Arabian oil fields when you really only had a four-hour stopover in the lounge at Dubai airport?" I countered.

"I never said I worked in the Middle East. People assume what they will. There's no benefit to me to over-clarify my work history," said the Super Manager.

"Teach me your ways."

◊◊◊

Every company I had worked for had an accountant or two who did some economic evaluations. Because *the corporation* was so enormous, it had a whole floor of business analysts (or BAs). The BAs had elaborate spreadsheets and spent their time evaluating options (even dumb options) to ensure that we were continually making optimum decisions. The BAs didn't really understand how powerful they were, even though their job was to do maths in order to inform management so that management could make great decisions. I don't think that there were many managers who would do anything unless it had been 'endorsed' by the BAs. The BAs for their part were pretty geeky. I guess that happens when you spend all your days staring into a spreadsheet. Trevor was the BA assigned to my project, which meant that I had to convince Trev that what we wanted to do was ok, so that he could do the magic numbers in order to convince the rest of the organisation. Trev was a pretty cool dude. He told me that *the corporation* was like the giant machine on tank tracks that slowly but surely rolled across the deserts of Tatooine selling droids (from Star Wars). You could stand behind the behemoth and push as hard as you could, and it didn't move any faster. Alternatively, you could stand in front of it and push back as hard as you could, and it wouldn't move any slower. Sooner or later, doesn't matter who you are, you get on board and watch the desert landscape roll by (at a constant speed).

Trev was never allowed to be a manager. I'm not sure if this was exclusively due to his inspirational metaphors.

After one year at the greatest company on earth, I got a letter from HR. It pointed out that I had in my salary package an allowance called "market premium." Basically, it was about a 10% uplift to my base pay because I was recruited in a boom time and the salary had to be competitive to get me. The letter asked me to acknowledge the market premium amount and further acknowledge that *the corporation* could elect to stop paying the market premium if market conditions in the future changed. I was aware that my contract included a salary item named market premium but was not aware that *the corporation* could choose to stop paying it. I checked my contract and the correspondence of the time leading up to my engagement. There was never any mention that the market premium was discretionary. I rang the lady who ran HR (later I found out that she was thought of as the grim reaper; sometimes you would hear whispered: "Death is on the floor" when she visited our building) and said:

"I'm confused, you've sent me this letter that you want me to sign twice and send back to you. It appears totally unnecessary in relation to my contract. In fact, if I were to sign it, I would be agreeing to a degradation of my rights under our contract."

"People don't usually read them," she replied.

"Well, surely you don't expect me to approach my contract with *the corporation* with any less rigour than the way I approach the contracts that I manage on behalf of *the corporation*?" I had thought of that line the day before and really enjoyed delivering it.

"Just sign it and send it back to me," she replied and hung up. Needless to say, I didn't sign it or send it back. I was, however, paid the full contract market premium rate for all the years that I worked there.

I was invited to a *Manager Once Removed* interview or MoR. This was something peculiar to *the corporation*. The boss of one's boss was known as one's MoR. Once a year, if one was assessed as worthy of such an honour, one was invited to have a formal meeting with one's MoR. There were even minutes recorded. My MoR was a guy named Daniel Sillcheski. We called him Silk-testis. I went in to his office and started off saying: "Before we get started, I just have to tell you how good it is to work for Angus. He's a really good boss and really gets the best out of me."

"That's not what this is about," Silk-testis smiled. It seemed he'd heard this before. "Your manager manages your work, but your manager-once-removed manages your career." I was pretty sure that my career was going to lead to other organisations outside *the corporation* but I didn't feel that it was the best time to tell Silk-testis that.

"In that case," thinking on my feet, "Angus is an awful boss, and he's borderline incompetent and you should sack him…and promote me." Another smile from Silk-testis. After that fantastic introduction, we went on to talk about my work history and career aspirations. He listened intently, gave some sage advice and took some notes. It was very encouraging to be given some attention in this way. I thought that with this kind of support from above, my career advancement was a certainty. As it turned out, Silk-testie did nothing for my career. Nine months later, he had moved on to another role, two years after that he left the business. I'm not sure where the meeting minutes had been filed. Five years later (after another four MoR interviews, each with a different guy, as the important people continually swapped places), I was still doing the same job.

◊◊◊

The "resources industry" (a collective noun to describe both 'mining' or 'oil and gas') relies on the discovery of "deposits." A deposit is a large quantity of high quality exploitable "resource" naturally occurring somewhere. For the mining industry, the resource is a "mineral." A mineral is a rock. The Good Lord placed most of the deposits in the world away from where people want to live. For example, oil is produced (drilled up out of the earth) in the coldest and the hottest parts of the globe – namely in the North Sea above Scotland (North Sea is code for Arctic Ocean because it makes it sound warmer than it is), and Arabia (no explanation needed).

Another bulk commodity extracted from the earth is iron ore. Of the world's ten largest iron ore mining centres at the time of writing, two are in the Brazilian jungle mountains, one is in inland South Africa and the other seven are in the Australian Pilbara region. The Pilbara is an enormous desert, distinguished from the rest of the Australian outback solely by the fact that all Pilbara dirt is at least 40% iron oxide. Iron oxide is commonly known as rust. Hundreds of millions of tonnes of rust are dug up from the Pilbara desert and put on ships bound for China. Most iron ore goes to China. Why China? Well, there are two possible explanations:

1. China has lots of big "smelters." (Smelters are industrial factories that use iron ore and carbon and heat to create "steel". Steel is much more useful than iron); or

2. China has a rust deficit that needs to be rectified.

The Pilbara is a desert. Most people up there say that they love it. I hate it. The temperature peaks at 50 degrees Celsius. Summer brings on daily lightning storms. Bushfires are a regular occurrence. Multiple cyclones hit the area every year. When it rains, it pours. And the fine red dust gets into everything! As one of my electrical engineers used to say: "Well, it's the Pilbara, so if it can go wrong, it will go wrong." It is a thoroughly inhospitable place. The exclusive attraction is that there's iron in them there hills. If it wasn't for the iron ore, no one would go there. Ever.

In order to get people to attend remote sites like this, the resources industry invented FIFO. FIFO stands for Fly-In-Fly-Out. So instead of living in a town with your family and working close by (like most people in the history of the world), you can leave your family (who live in a city or town) and you can fly to a remote site and live there without them. Of course, after somewhere between one and four weeks at work, you can then fly back home to your family for a week off. (The week off is usually spent being tired and grumpy for two days, then going out with your friends for two days, then spending the next three days trying to get out of doing the list of jobs that your wife has compiled for you.) In return for this ridiculous lifestyle, FIFO pays really well (both in real money and in airline frequent flyer points). It's a really good thing to do for a couple of years to build up the bank account, but it's a really bad thing to do forever. Most people start working FIFO with a plan that generally has the following steps:

1.Work a few years;

2.Save up some cash;

3.Get out of FIFO; and

4.Buy a house.

However, the FIFO reality for many is generally more like this six step process:

1. Work a few years;
2. Save up some cash;
3. Spend it all on big toys and a massive overseas bender holiday;
4. Repeat steps 1 to 3 until you get married then move onto step 5;
5. Buy a house that's too expensive (for your wife and kids to live in) that you can't afford unless you keep working FIFO; and
6. Keep working FIFO

I flew to the Pilbara every week for four years. I usually took the site charter plane that went directly into the mine site, landing on a private airstrip. The planes that we chartered were F100s; the worst jet planes in the sky. As well as being the worst, they were also the cheapest jet planes in the sky. The F stands for Fokker (the aeroplane manufacturer) and 100 stands for the number of seats (and consequently passengers). F100s were designed in the 1980s and as well as being the worst and the cheapest, they were also the oldest jet planes in the sky. There are no entertainment systems in F100s. When I started doing FIFO flights, the company that owned the charter planes (SkyWest) had a blue logo and the planes were painted with a blue tail. SkyWest was bought out by Virgin Australia, who set about rebranding the F100s and painting the tails red. I always felt better in a red painted plane; it gave me the impression that the plane was newer, or at least new enough to be worth repainting. Eventually, I guessed, all the blue planes would be cannibalised for spare parts to fix the red planes. Then, the red planes would start eating each other until there was only one left. Maybe then planes made by Boeing or Airbus might be old enough to be used as FIFO charter planes.

Being a mechanical engineer, my training led me to wonder:

o   How many fatigue cycles these wings had been through over the decades (a little knowledge is dangerous);

o   Whether all the maintenance had been done (or if it was like my car where I am convinced that the mechanics just stamp the book without opening the bonnet); and

o   How much advancement in aerodynamics had occurred in the last 40 years (I'm sure that modern planes have longer wings – the F100s seem to bounce around all over the place).

This musing, combined with enough education to make me think that I understand the basics of aerodynamics, led me to develop an anxiety of flying. I don't experience any panic when flying on modern planes that have the names of companies that still exist written on the side.

There are two kinds of charter planes. There are full ones and empty ones. Full ones are more common and resemble a jail bus (I've never been in a prisoner transport bus but I have seen them in movies). On a full charter, you hope that you're not sitting next to Bubba the driller who has been on the drill rig all day and didn't have time to have a shower before flying out.

There are a few largely empty charter planes. This phenomenon occurs when a plane flies to site full (to suit the site roster) and then has to return empty (to suit the site roster). If you were far enough up the tree (manager level, like me), you could tailor your days on site to suit flying on empty planes (and I did). Regularly I got on a plane where there were more hostesses than passengers. Once, instead of watching the safety demonstration, the passengers (two of us) got to do the safety demonstration while being critiqued by the crew. We nailed it!

Nothing boards as quickly as a charter plane. The passengers are special (they all wear yellow or orange shirts and know exactly what to do and how much hand luggage they are allowed to take on board) – there are never any passenger caused delays. Even the terminal is special (very basic construction and all the planes are F100s).

The most painful flight I had was flying back home on the regular two-hour trip. This time it was complicated because there was a thunderstorm at our destination airport, and after two hours of flying it didn't look like we were anywhere near home. At this point the pilot told us about the thunder storm and that we had been diverted to a regional airport. When we got to the regional airport, we lined up our plane next to two others. Then the pilot told us that we needed to re-fuel because having flown all this extra way, we didn't have enough fuel to get home. We had to remain on the plane, preferably seated in our seats for probably the next hour. Did I mention that these planes do not have any entertainment system? He also said, somewhat cryptically, that since we couldn't clean the toilets there, we needed to be considerate while we were waiting. Translation: Don't take a dump until we get home because if you fill up the toilet, it's going to stink the whole plane out!

Out of my window I could see the fuel truck filling up the first plane in the queue. It takes a long time to pump a plane full of fuel. But after about half an hour, it moved on to the second plane in line. The first plane left the queue, and presumably the airport, to make its way home. The second plane took the best part of an hour to fill up, and eventually it too departed and the fuel truck moved on to our plane. After about ten minutes, the fuel truck unhooked and drove away. Wow, I thought, that was quick! But how is it possible that our plane only needed a little top up? This didn't make sense to me. But all was revealed when the pilot came back on the microphone for his third message: "Hi folks, unfortunately the fuel truck has run out of fuel and has to go and fill up before coming back to finish refuelling our plane. It will take about an hour for the truck to get back here. Please be patient."

All up we were on the tarmac for over three hours which, when added to the extra flying time, meant it took over six hours for this normal two-hour commute. The lady next to me was really angry but I had to laugh – what a comedy of errors: you couldn't have scripted this string of fiascos any better (or worse) if you had tried.

◊◊◊

In the evenings, there are two options in camp. You can go to the gym or you can go to the 'wet mess' (sometimes referred to as the Tavern. It is essentially a pub in the mining camp, but is not to be confused with the 'dry mess', which is a food hall). So, you could pump iron or you could drink beer. Or, if you timed your evening properly, you could fit both in. The gym was open 24 hours a day and was popular, but the wet mess was even more popular.

Tuesday nights I spent on site at the wet mess enjoying $3 beers with my crew (a beer cost about $10 in a normal pub at the time). The surveyor on site called me 'Cheap Beers Tuesday' - or just 'CBT' for short. This was where the real management happened. This is where CBT found out what was really going on in the site. I built relationships with everyone on my team: the site management, the engineers and the contractor personnel. I told every group that I would be in the wet mess on Tuesday nights, so come and see me if you want to talk about anything. Occasionally, someone would take me up this this offer, but usually it was just ad hoc good times.

The rule of thumb on mining camps is: 9 before 9. I'll explain: drink no more than 9 beers before 9pm and you would still blow zeros in the morning. (Everyone was breath tested every morning and if your breath alcohol level was higher than 0.00 on the breathalyser, you were sacked. 'Blowing numbers' was the idiom used for not blowing zeros. Clever – any number, in any position, led to the same outcome: sacked.) The wet mess only opened at 6:30pm and I couldn't imagine consuming nine beers in two and a half hours. I reckon that I'd still be drunk the next lunchtime. Needless to say, I never tested the rule to find out just how many beers would translate into blowing numbers in the morning.)

The bad thing about living in a mining camp was that you had to live in a box, called a donga. Construction dongas were the smallest kind of dongas. Ten foot by ten foot rooms in which you could stuff a bed, a tv, a fridge, a small wardrobe, a shower, a toilet, an air conditioning unit and, at night time, a person. The good thing about living in a mining camp was that you got a box to yourself. I could never sleep well in camp. It didn't matter which camp (mind you they're all as bad as each other; even the good ones are bad). I don't know how the guys did it, living up there for four weeks straight. The typical construction roster was called 4 and 1, i.e. 4 weeks on and 1 week off. Twenty-eight days straight living in camp before seven days at home with your family. I couldn't do it. I realised that I actually liked my family and spending time with my lovely wife and kids. Being away for just one night a week was taxing enough for me. Luckily, I was so important that I could just spend about twenty hours on site each week (which I would describe as two days if ever asked by my management).

One day my darling wife, Bella, ironed my site shirt and pants. The shirt was yellow, with my name on the right breast and the name of *the corporation* on the left breast. It had beautifully crisp pleats. I felt trapped. I could either ask my darling not to iron my site shirt in future and risk her feeling that I was ungrateful, or I could wear the shiniest, brightest shirt to work and get teased by the workers. It was bad enough that my shirt was never dirty; I didn't have to rub their noses in it by having pleats too. The guys on site referred to a clean shirt as a "manager level shirt." (Having 'manger level' – clean – boots is considered even worse).

The trick to spending a night or two on site is to make sure that you don't pack anything more than you need. I adopted a minimalistic approach to packing. If I brought an item home that was still clean, then I had failed. This approach led to fewer and fewer items through a continuous improvement philosophy. In the end, I only ever took one work shirt to site regardless of how many days I was staying. Before you imagine the stench of a week-long worn shirt in a 40-degree, dusty environment, let me explain the protocol. At the end of the work day, you could wear your shirt in the shower, and before washing the red dust off your body, you could spend a couple of minutes washing the red dust off the shirt itself. You could then hang out your wrung out shirt outside your donga and go to dinner. For the nine months of summer every year in the Pilbara, your shirt would be dry by the time you got back. During the short-lived cooler months, there was one more step to be undertaken. After wringing out your shirt by hand, you could lay it on a towel, roll them both up together, then wring out the whole towel wrap again. The shirt is almost dry enough to wear straight after this treatment. Hanging it overnight finishes the job.

The next morning, I would be up at 5am to leave the camp at 5:30am. The FIFO residents typically arise in the 4s and go and have breakfast. Bacon and eggs at 4am. Delicious. Occasionally, a new arrival at camp might ask what time the mess opened. The answer is of course, if you're awake, the mess will be open. I would normally skip breakfast because I felt that before 5am, sleep was more valuable than food. As the day wore on, the equation changed and by about 8am, food became the priority again.

Contractor pre-start meetings occurred at 6am. Pre-starts are exactly that; a get together before you start work. The supervisors would talk to the guys about the tasks for the day, the weather conditions and anything else that was relevant before they started work. I viewed this gathering as a captive audience on which to practice my public speaking. I would cycle around the contractor pre-starts (a different contractor group each week) and deliver the same message:

"I just want to make my expectations clear: I don't want anyone to ever put themselves at risk on this job. Don't think that you're doing me or *the corporation* a favour by taking any shortcuts that put you or your buddies at risk. We won't thank you for it; in fact, we'll punish you if we find out. Take the time you need to do the job properly and safely the first time. I want everyone working on this job to leave the job *better* than when they started. Work out what that means for you personally. Come and see me in the wet mess on Tuesday nights." It's amazing how giving a damn about people really works. Traditionally, no-one gave a damn about the guys who work on remote construction sites, so even if you only gave a little damn it was still a lot more than they were used to.

The key to speaking to groups is really quite simple. Forget any management training that you have had, forget any expectations that you may have built up about how to speak to a group, and forget any pre-conceived notions of how to act like a leader. Pretend that the guys in the audience are friends, and simply speak like you normally would to such friends. Honesty goes a long way. We are human beings who are very good at verbal communication. After all, we've been doing it for thousands of years. People don't really respond to *what* you say. People respond to *what you are being* when you say it. It's mostly in the non-verbals. We humans know when we are being lied to (so don't do that).

Occasionally, I was required to be on site for a couple of days in a row. The most memorable time was in the lead up to Christmas one year when all the PMs were instructed to go to site for at least three days a week to make sure that there were no injuries before the break. My opinion was simpler – I was an advocate of all PMs closing their eyes and crossing their fingers as this would probably yield the same result. But it didn't seem the right time to express my thoughts, so I mobilised myself to my site for three days a week. I interpreted three days as two nights on site. So I could fly up on a Monday afternoon and back on a Wednesday morning and satisfy the three days on site rule. There was a benefit to the guys working on site to see that their PM gave a stuff enough to turn up for a few days, even if I had nothing to do. And boy did I have nothing to do.

Having set the site up to run without me needing to be there, I was instantly bored. I still went to pre-starts and inspected the odd work front, but for most of the day I was busy trying to stay out of the way and not upset the smooth running of the site. Two consecutive 5am starts was rough. I don't know how the workers cope doing this for four weeks at a time. There was an afternoon when I went into a meeting room, shut the door, and lay down on the floor under the pretence of stretching my back. I awoke when one of my site engineers opened the door, looked down at me and pulled a funny face. He then had the good sense to leave quietly, shutting the door behind him without saying a word. He set himself up for promotion that day. Nothing like a mid-arvo nanna-nap and colleagues who leave you alone. Most of the rest of those long days on site were used to write this book.

There was one colossal balls-up that occurred on my project. My contractor cut the rails out from underneath an operating rail-mounted stacker. The stacker is a machine that travels up and down rails about a kilometre long, stacking ore (dirt) into large piles. The piles are then later reclaimed (collected by a different machine) and put on a train. Part of the project was to modify the rails. In an amazing lack of good judgement and without the proper permits, some guys went and prematurely cut out the wrong rails while the stacker was down the far end. Losing a stacker is very bad because if you can't stack, the whole mine stops running and you lose mega dollars every hour. The next morning at 9:08am the stacker automatically drove itself off the new end of the rails. I know it was at 9:08am because at 9:00am every Tuesday we had a weekly progress meeting in head office with project leadership team. At 9:09am three phones that were sitting on the meeting room table went off simultaneously. Before I picked up my phone, I knew something big was going down…and big didn't mean good.

I was told what had happened with the stacker, and that site operators were justifiably furious. I rang my boss immediately.

He was known (behind his back) as the Primary Crusher or PC for short. This is a mining joke because most mines have a large primary crusher that turns big rocks into smaller rocks. You don't get a nickname like the Primary Crusher unless you have a certain tactless, abrasive way with people. He was widely regarded as a bastard who trusted no-one. Even his own Project Managers knew that they were the "scapegoats in waiting." When the time came, the PC would sacrifice any one of us for his own career. I had the displeasure of reporting to the PC for about a year.

Ring, ring…I was hoping that it would go to voicemail….ring, ring… "Hello." Damn!

I told him what I knew and that I would be on site in three hours' time. I rang him back about ten minutes later as new information came to hand. Then again, half an hour after that. Again, an hour later when I was at the airport, and two hours after that when I was looking at the machine with the wheels in the dirt. And twice more that night. He knew what I knew as soon as I knew it.

I found out years later, when the big boss' secretary told me, that the Primary Crusher didn't tell his boss (the big boss) until 4pm that afternoon (after I told him that we had a plan to fix the rails on night shift and that the stacker would be running again the next day). The secretary said that the big boss had asked her who does this Jack think he is, not reporting the stacker rails incident until eight hours after it happened. So not only had the PC not told his boss for hours, but when he did, he blamed his tardy reporting on me. I didn't like working for him.

The next morning the birds were singing, the bees were buzzing, and more importantly, the stacker was stacking. And the investigation commenced. Investigating incidents can be more of an inquisition than a genuine fact-finding mission. The intent is to find out what actually happened and then to set in place controls to make sure that a similar incident could not happen again in the future. It can take a week or more depending on the complexity of the incident. Unfortunately, there is a tendency for investigations to become charades that serve only to identify who to blame and then sack. I was determined to make sure that this investigation was done properly. I even stayed on site for three days in a row.

The investigation still had a day or so to go when I returned to head office on the Friday morning. The PC had a large corner office. The view looked out over the streetscape with a church in the foreground. A small replica samurai sword was in a little glass case on the bench in front of the window. I was wondering if the PC had ever used the little samurai sword as I filled him in on all the details of the incident, the recovery and the investigation. When I had finished my report, he looked me in the eye and held up his right index finger saying slowly: "one from the contractor," at this point he raised his right middle finger to accompany his index finger, "and one from the engineer by the end of the day."

I was initially confused, probably because he had not used a proper sentence, just jammed two phrases together without a verb.

Or maybe because of the totally unreasonable, unfair and uncharitable implications of his statement. My mind was racing, trying to work out exactly what he was saying. Time slowed down.

…Is he telling me to arbitrarily sack two guys today…regardless of fault or intent?…

…Is there any other possible way to interpret it?…

…Why am I having trouble understanding this very simple execution order?

…The mini samurai sword is shining in the sunlight…

I was really shaken, but I did my best to keep an ordered, professional tone saying:

"I think it would be prudent at this stage to complete the investigation first. Once we know exactly what happened, we can take actions to ensure that it can't re-occur. This may include removing people from site if that's the appropriate course of action." I was essentially trying to buy time, trying to just get out of his office without promising anything. Tomorrow was the weekend. By Monday things would be two more days in the past. Time heals all wounds (except the ones where you can bleed out, or ones that are badly infected with gangrene). Maybe this would take more than two days, but two days would be better than none.

The Primary Crusher replied by holding up his two fingers again, one after the other: "One from the contractor and one from the engineer, by the end of the day."

There was a minute's silence. So this is whey they call him the Primary Crusher.

"I'll get back to you." I got up and left.

As I was walking back to my office, I was extremely troubled. I was certainly not going to arbitrarily take two people's jobs away as I had been instructed. If the investigation had been completed identifying people who were clearly at fault and who had no remorse and were obviously going to re-offend, then that would be a different story. I was trapped since my morals didn't allow me to follow the clear verbal instruction from the PC. I didn't know how this was going to end. It did occur to me that the logical conclusion, considering my inability to comply, may well be that one Project

Manager could be leaving the project by the end of the day. Why on earth did he have a baby samurai sword in his office?

I was walking slowly and praying fast.

I was disturbed. I had just been instructed (by the biggest manager-bully in *the corporation*) to sack two guys arbitrarily and instantly. I couldn't do it; this was wrong. I thought to myself that since I was not going to comply with the PC's instruction that this was probably my last day at *the corporation*. I imagined my next interaction with him – informing him that I was not going to do the reprehensible action that he had instructed me to carry out. He would then, no doubt, be the architect of my demise. It was only a matter of time (this is, after all, the role of the scapegoat in waiting).

I phoned the Engineer's Project Manager, Edwards, and explained that we were under pressure to show some actions out of this investigation. And that the pressure was on me, not him, and I wanted a quality outcome rather than a knee jerk reaction.

Edwards told me that he could no longer trust his supervisor to make good decisions since he'd denied any responsibility (even though he was supervising the work). He had repeatedly blamed everyone else around him. Edwards had decided to remove the supervisor from site that day which he had already actioned.

"Well, Edwards, I support that." My support wasn't exactly necessary.

Then I rang the boss of the contracting company, Ivan, and had the same conversation. Ivan told me that they were doing their own investigation in the background by a different method. Although my investigation was not quite complete, their process had determined that one of Ivan's supervisors *and* one of his engineers was highly at fault, and they would both be leaving the project (and the company) the next day.

"Well, Ivan, I support that." Again my 'support' was largely tokenistic.

I was able to go back to see the Primary Crusher late that afternoon to inform him of the 'good' news. One from the engineer and two from the contractor would be leaving the project in the next

12 hours. He didn't want to know any details, but just waved me away with a nod that said, 'Well done, lad'. I felt a bit dirty.

The next week, when I was in my weekly one-on-one meeting with the PC, he told me: "Jack, I have to tell you that the business is very appreciative of the actions that you took last week."

Is there no end to this man's manipulation? "But weren't they your actio…thank you…I guess." I got another approving nod, as I got out of his office as fast as I could.

Retrospectively, it is very interesting to note that there is no written record of the Primary Crusher's instruction to me. So, in a way, the PC was right in identifying the actions as mine since there is no evidence that records his involvement in this affair. Apart from this book that is. And the hundreds of people that I've told in pubs.

This episode cemented in my mind that the Primary Crusher was not someone I wanted to work for, be friends with, or even know.

◊◊◊

I was once invited to a presentation to the Japanese Joint Venture Partners. A Joint Venture (or JV) is where you share ownership of an asset (in this case a mine) among two or more companies. Since only one company can operate the asset, one party will the operator and the other(s) will be non-operational or 'silent' partners. They always start with good intentions, but after some years or decades, the silent partners eventually start to mistrust the operating partner and the operating partner resents the silent partners 'control' of capital expenditure purse strings.

(Many years later I was told by a retired JV manager that the first rule of JVs is: 'Don't have them.' The silent partner is an extra party from whom you need to get approval (in addition to your own management) whenever you need money from them. And anything that you want or need to do requires some of their money. Every time that you ask them for money, they hold off giving their approval until they've extracted something extra from your *corporation*. Sometimes it's just information they're after but it could be a renegotiation of the profit-sharing arrangement. You never know until sometime after you've asked for the money.)

I flew with a contingent of more important managers than me to see the Japanese at a neutral venue roughly half way between us and Japan. We arrived in the late afternoon and hurried straight to dinner where a large contingent of the Japanese was waiting for us. When I say waiting, they had already commenced drinking saki although their pace appeared to increase markedly when we arrived.

This arrangement confused me a bit. I was much more used to the European method, where you arrive, make the presentations, do the deal, and then start drinking together. But this Japanese method seemed to be the opposite. Spend all night drinking first, then get up the next day hungover and try to make your presentation coherent.

When I collapsed into my bed at about one o'clock in the morning, I had two thoughts. The first was that our presentations start in eight hours and I wondered how much my head would hurt in the morning. The second was that I was sure that all the graduate Japanese engineers would probably have to write notes about everything they heard at dinner before they were allowed to go to sleep. I was glad that I wasn't a graduate or Japanese.

The presentations went well, I only had two slides at the back end to talk to. They were much more interested in the financial deal than how we were going to construct it. I think I was only really there to give some credibility to our construction prowess. It certainly wasn't for my saki drinking skills.

In the years since, periodically I get asked whether I'm familiar with this JV. I reply: "it was some years ago now, but yes, I drank the saki," with a knowing smile and slow nod. Occasionally I find someone who smiles and nods back.

◊◊◊

Rusty was the Engineering Manager for one of the consultant companies who worked exclusively on our projects and lived within our building. This meant that he worked on multiple projects; not really dedicated to any. He was there to 'manage' the engineering across the projects. This really meant managing the people doing the engineering, i.e. minimising engineering cock-ups while maximising chargeable hours. In short, Rusty was what we called "an overhead." Charles was Rusty's boss – his General Manager. Charles was even more of an overhead than Rusty. He was totally focussed on maximising chargeable hours and didn't really care about minimising engineering cock-ups.

One day Rusty came to me with a form for activating payment of Charles's retention bonus. The retention bonus was quite a ridiculous system with the single aim of dribbling out overpayments to the members of the consultant companies, effectively drugging them with money. This kept them dumb and lazy – but present.

Rusty asked me to sign Charles's form saying that Rebecca from accounts had checked the hours and effectively all but approved it. All that was needed was for all the PMs to sign for the hours that were allocated to each project. He showed me the form that already had seven signatures from my colleague Project Managers, approving their parts of the bonus payment. I looked at the form and I could feel a disagreement coming on. I didn't know Rusty very well, so I didn't know how this disagreement was going to end, but in short, there was no way I was going to sign that form.

"Rusty, there is nothing that you can say to convince me that this retention bonus system was designed to retain your *General Manager* in the program. It's designed for the workers, not for those like Charles who are paid in the top 1% of national wages. In fact, the idea of paying Charles enough to buy a Ferrari as a retention bonus after we've already funded two Ferraris for him this year is pretty disgusting. I am having trouble seeing this as anything but greed on the part of Charles, which would really compromise his character, probably to the point where I'd have trouble trusting his professional judgement. If Rebecca has 'approved it' then maybe you should get her to sign it." I knew full well that Rebecca had no signing authority within *the corporation*.

"So, no, I won't sign it. My advice would be to cut your losses and submit the form with the seven signatures that you've already got. If you want the eighth one, then you could escalate to my boss, Angus, but I'll warn you that if you do that, it may not end the way you'd like it to."

Rusty looked at me, silent for a moment, then with a slight nod of his head said: "got it," and turned to leave. That was such an anticlimax. I was expecting at least some resistance. Maybe a bit of push back. What won the day? Was it my eloquence? The Ferrari imagery used to illustrate my point? Or maybe Rusty actually agreed that this was just pure greed and wrong. Before he got out the door, I thought that I should clear up an unresolved matter that would ensure that Rusty and I never had to have this conversation again.

"Rusty, just so you know, when you bring me a form like that with your name on the top, I won't sign it either."

"No problem."

◊◊◊

Many times, we were instructed to roll out a new system (that normally replaced an old system that everyone knew how to use). The PC would have presented this by extolling the virtues of the new system and stating how it was superior to the old system, before closing by explaining why it was good for us to use the new system. No-one really believed that the new system was a benefit. Mutsy would go about this very differently. He'd say: "This is the new system that we're getting forced to use. We can talk about why you don't like it, if you want, but you're going to have to use it anyway." Everyone knew where they stood. We all liked Mutsy. He treated us like adults and told us the truth.

If you ever find yourself high up enough in the management echelons to be making decisions about new and existing systems, remember that every time that you add a new one, you have to remove an old one. (It's just like your wife's shoes. For every new pair, an old pair must be jettisoned to make room. Otherwise, the shoes will spill out of the cupboard and create trip hazards. The irony of shoes as trip hazards should be avoided at all costs.)

Most *corporations* do not use this simple rule. I know it's surprising, but how could multi-national organisations miss such an elementary mathematical equation? But they do. Typically, someone high up finds something that another company is doing and copies it. Occasionally, someone high up may have their own independent thoughts and come up with something new all by themselves, but examples of this are rare. New things are usually copies of other companies' practices. This new system is then thrust on the employees, usually with no knowledge that there is already an equivalent system being used. And all of a sudden you find yourself having to comply with two (or even three) systems all doing the same fundamental function.

There is only one way to test if a system or procedure is obsolete. Stop doing it and see if you get into trouble. I told my team to stop publishing monthly reports at one point. Monthly reports are a cornerstone of any engineering project management system. They should capture the current status, cost forecasting, schedule predictions, issues, and any other important stuff. At *the corporation*, it was standard practice to produce two monthly reports. The first was called a Flash Report. Flash because it was very brief. It consisted of four graphs on a single page so that at a glance, the reader could see the progress and status of the project. The Flash Report was a very good thing. It was distributed widely. The graphs for each project were combined together into master graphs that allowed portfolio reporting. But most importantly, the Flash Report was read…the mark of a truly useful report.

The second monthly report was simply called a Monthly Report. But there was nothing else simple about it. It was a verbose document of many pages, written by many people, collated by yet others. It was written but never read. I didn't read it. My boss didn't read it. In fact, as far as I could tell, no-one ever read it (except for graduate lawyers if you ended up in court with your contractor). I got away with not producing monthly reports for eighteen months before my non-compliance was discovered – by accident. The big boss' secretary asked for my most recent monthly report, not because she or her boss wanted to read it, but so that it could be used as an example of *the corporation's* best practice for a new PM. I told her that she probably didn't want to use my most recent monthly report because it was eighteen months old. She was kind to me and said that she wouldn't tell on me, and I promised to start producing them again.

◊◊◊

My projects all went really well. The team that I had put together and built into a high performing team (through the power of trusting relationships) delivered early, under budget, with no injuries every time. It wasn't because of me, it was because the guys all knew what their unique jobs were, and they trusted the guys next to them (just like a good rugby team). Once we went more than 30% under budget on a billion-dollar job. I thought that this would have been a good thing, but we had to write a board paper to explain why we hadn't spent enough money and sort of get permission to underspend. I wrote the first draft of the board paper and told the truth: the guys who did the estimate before my team got given the project were incompetent. It was rewritten by others who were more attuned to *the corporation's* expectations.

The team had developed high expectations of their own performance. A telling characteristic was that no-one accepted delays. From the top to the bottom, everyone completed their work on time. On a number of occasions, suppliers would ring up and tell us that they would be late in delivering their equipment – always for reasons outside their control. The conversation would always go the same way:

"You can't be late. You're contracted to deliver on time."

"It's not my fault, we can't because…" there were always different excuse details.

"I'm going to ask you three questions: What do *you* need to do to get back on schedule? What do you need *me* to do to get you back on schedule? And what do you need me to get *my senior management* to do to get you back on schedule? For example, does my boss need to ring your subcontractor?"

"I'll get back to you." Hang up.

Generally, the suppliers just needed to know that we were serious about schedules and would not accept late delivery. Once they knew this, they would find ways of delivering on time. We would joke about doing *whatever it takes* to get the equipment delivered on time. It was pretty rare that we actually had to get involved to assist, but it did happen. Our switchboard supplier told us that the factory in Germany was late manufacturing our switchboards, and therefore they were going to be late. The engineer who met with them replied that they couldn't be late, because his Project Manager was personally interested in this package and had booked airfares to travel to Germany to attend the FAT (Factory Acceptance Testing – testing that occurs in the factory by the customer's representatives before the equipment is released). *Whatever it takes.*

They scurried off to magically get the build back on track. The engineer came to see me after the meeting. Had they recovered their schedule? "I think that they'll find a way to recover back to the contract schedule. And I sort of told them that you were going to Germany to attend the FAT. Can you get approval for that?" he asked. International travel was a tightly guarded activity at *the corporation*. It was for extremely important people only (who often complained about having to do it so much). However, at my level, I needed to get three signatures to approve this. I explained the situation to the Primary Crusher, who said no. A week later, I asked him again, and he said no. In all, I think I asked the PC six times until he finally said 'ok'. Maybe he just wanted me to stop asking.

I was off to Germany for a week of electrical testing. *Whatever it takes.* After all, it had been years since I had had a European junket for work. I was careful to take an electrical engineer with me – after all, it seemed prudent to have someone alongside me who actually knew electrical stuff. We spent a week in Frankfurt. It was autumn. The weather was nice. The food was fantastic. I bought a wool suit (on special). Oh, and we went to a switchboard workshop. In all, four of us flew over to Germany for the FAT; me and my electrical engineer, and two guys from the supplier's organisation. We went into the factory for the first day of testing, which was scheduled to take four days. If you're not trained in electrical engineering, this kind of testing is the most boring thing that you can do. It is worse than watching grass grow or watching test cricket (although, at least in test cricket you can understand the rules of the game). And as the client representative, you can't sleep in a meeting room – it would be a very bad look. The other guys were feeling the same way, so we left the electrical guy to witness the testing and the rest of us took off to be tourists. We saw the sights of Frankfurt, we went running along the river banks, we hit the shops, and we even took a train to Dusseldorf to inspect a gearbox factory run by the same company (which was slightly more interesting than electrical stuff). Whilst the testing was being dutifully witnessed and accepted, we ate too much during the day, drank too much at night, and spent the afternoons in the Turkish baths under the hotel. If sweating in a sauna with the supplier reps was the price of on-time delivery, I was the right man for the job. *Whatever it takes.*

◊◊◊

Leadership was the flavour of the month (every month) at *the corporation.* The bosses talked about leadership often. Better leadership was the answer to everything. Ironically, I didn't think that those espousing the superlative values of more leadership were actually very good leaders. The good bosses I'd worked for (Angus, Mutsy, Pedro) were all a little weird. They didn't fit into the carefully choreographed expectations of leaders. They were first and foremost *real* people. They drew people to them, they created an identity for their team, and they had charisma. Or at least they gave a damn about me. The leaders at *the corporation* fit a certain mould. They were the grey men. The grey man manager was characterised as follows:

- o He wore the same suit as his colleague managers (often in the same size) – it was charcoal;
- o He had never made a mistake (this is best achieved by never making a decision);

o He was upwards focussed, always asking himself how he could enhance himself in the eyes of his boss (usually to the detriment of his subordinates); and

o He was compliant, favouring the tried-and-true methods of the past rather than trying anything different (see point 2 above).

When someone was appointed to a senior position, he or she was systematically 'greyed-out.' It must have been a formal training program that polished most of the personality out of a man. For those who were a close fit for the mould, this process only took a few weeks. Some people took years to get fully greyified.

I set about being the best leader that I could be. I got to know the people in my team, where they came from, what motivated them, where they wanted to go. I set myself the aim of being the manager that people wanted to work with. To be the boss that people wanted to be directed by. The leader that people would follow...anywhere. Dwight D Eisenhower (one of the Presidents of the USA) famously wrote: *motivation is the art of getting people to do what you want them to do because they want to do it.* Have a think about that. Quite different to manipulation. How indeed do you perfect the art of motivation? Well, it's all about relationships. I recognised that I would go out of my way for Angus because of our relationship. And I vowed to never let myself be polished grey.

◊◊◊

In the world of engineering mega-projects, there are basically two ways you can work as a project manager. The first way is to manage a construction project. This may seem a bit obvious. Managing construction projects means being exclusively responsible for the:

o cost;
o progress;
o quality; and
o safety of everyone working on the site.

All in all, this is very onerous. The budgets are enormous and you need a team of people to just keep track of it, or you need a team of very good people to manage it. Mutsy used to tell us that the key to good project management is to make sure that we get enough money approved – you don't get yelled at for giving money back.

Likewise, the schedule can be years long with multiple critical paths. Without good schedulers and planners, you simply have no idea what is going on. Mutsy also told us to make sure that we get enough time approved. Going overtime is not normally as bad as going over budget, but you still get yelled at.

Managing quality is a mysterious business. We all say that we do it, but really, we don't know what it means. If the plant you build works, then it is to quality. If it doesn't work, I guess that it's not up to quality. Mutsy's solution: make sure that you have enough money and enough time approved so that you can fix any unforeseen problems without having to ask for more money or more time.

Managing safety on the other hand is an art rather than a science. Good safety performance only comes about if every worker on the site makes continuously good decisions all day every day. To do this you need to be able to influence all the workers. And I mean ALL the workers, not just the managers, superintendents and important people, but everyone down to the lowliest labourers and cleaners. The bigger the project is, the harder this gets. My rule of thumb is that a Roman Centurion led 100 men (that's how he got his title of Centurion). It's not an accident that the Roman Army was organised into units of 100. A human brain has the capacity to know 100 people individually. So, a single Centurion knew all his men personally and could influence them all to move towards a single objective. If you're lucky enough to have a small project with only three hundred people on site, then you only need a small team of three to five excellent leaders to know and influence them all. On the other hand, if your site has 5,000 workers on it, you need a whole organisation of excellent leaders to keep the ship running in the right direction.

And despite having employed good leaders (who treated the workers as people), and having set up excellent safety systems on site,

I was always a little "on edge." I knew that at any moment, the phone could ring to say that someone had hurt himself, and by the magic of accountability, it would be my fault.

Once I came out of a movie theatre in the evening and turned my phone on to find a voice mail message saying that someone had cut his finger on my site and had a stitch at the local hospital. In normal life this wouldn't be a big deal. But at *the corporation*, it is much worse than it seems. According the 'rules' for classifying injuries, if the doctors use tape and glue to fix the injury this would be a 'First Aid Case' (or FAC). An FAC is good because it doesn't count as 'an injury.' This means that it doesn't hit the books and contribute the total number of injuries that eventually contributes to everyone's bonus payments. However, if the doctors use a single stitch to fix the guys hand, then this will be classified as a 'Medical Treatment Case' (or MTC). MTCs are bad because they do count as injuries and will attract attention from all of management who are focussed on protecting their bonus. Of course, the best way to not have injuries is to not have people injuring themselves - but you can't really 'control' that. I always told my site management team to make sure that anyone who gets hurt gets the best medical care available regardless of the 'classification.' We're not going to protect our injury classification at the expense of an injured worker. Anyway, when I got the voice message, I knew that I needed to report the 'incident' to my boss which would start a snowball of correspondence over the next week or so as every manager above me will try to get this re-classified into a FAC.

Managing projects is hard work – even if you're working for the greatest company on earth. Mutsy once told me that despite being the greatest company on earth, we were really just a bunch of dudes trying to work out how to do our jobs…just like everyone else. He never got promoted again.

The other way you can work as a Project Manager is to manage a study. For years before a construction project starts, people do studies. Depending on *the corporation*, the studies might be called things like: Order of Magnitude Study, Pre-Feasibility Study, Front-End-Loading Studies, Feasibility Study or Early Engineering.

It would typically take a minimum of three years to move through enough study phases (advancing the proposed project just enough at every phase) to get a project sufficiently defined to get it approved.

Managing a study is code for getting paid full rates to 'pretend' that you're doing a project instead of actually doing a project. It helps to have people on the team who have delivered actual projects because they are better at pretending that they're delivering a project than people who have only worked on studies and don't even know what they're pretending to do.

The best thing about managing a study is that you never get the phone call telling you that it's your fault that someone has hurt himself (people rarely get hurt pretending to do work).

At *the corporation*, we would have a study meeting every month. This consisted of about 20 people sitting around a large table reporting to the study manager about how little progress they had made in the last month. At a study meeting, all facets of the study are represented:

o   a bunch of approvals people (who all do different bits of government and environmental approvals);
o   the geology representative;
o   the geotechnical representative;
o   the hydro-geology representative;
o   the mine planning representative;
o   the metallurgy representative; and
o   me – the engineering guy.

It always seemed a little strange to me that there was only one engineer at these meetings. After all, when the project was finally ready to be approved, all the guys at the study meeting disappeared (on to the next study) and the engineering team delivered the project. But on the other hand, if my team came to the study meeting, it would have hampered my ability to make commitments on the spot in the meeting (the actual engineers always wanted actual data before making decisions).

During my first study meeting I pointed to the geologist and the mine planning reps. I asked them to explain to me how their jobs were different. They didn't answer but I don't think that they forgot me either.

About a year later, as this study was drawing to a close, the geologist when reporting his progress made the statement: "…the models are stranded…" My ears pricked up from my geoscience slumber. I guess that this meant that the geological model that he had spent the last year developing had got to a point that they could do something called strands. It was obvious from the context that this was a step forward. I couldn't help myself as I interrupted him from the far end of the long table:

"Um, excuse me," I raised my hand.

"Yeah," he looked up, a little confused. Evidently, he was more used to me day-dreaming during his updates.

"Did you just say that *the models are stranded?*"

"Yep, that's right," he was pleased to report this progress twice in a single meeting.

"Well, we should save them!"

Silence.

"All I can think of is a desert island with a bunch of stranded models in hula-skirts, maybe from a ship-wreck."

More silence.

I recognised the look called: 'how on earth did this guy get a job as a manager.' I'm pretty sure none of them forgot me after that.

◊◊◊

One lunch time, I took a call from Angus' secretary who asked if I could come up and see him straight away. I replied that I could re-direct the efforts of my entire team to suit his current whim. She simply said: "Thankyou." So, I went straight up and about two minutes later I walked into his office on the 34th floor. As I entered the room, he threw ten stapled pages at me, asking: "What is that?"

"Umm….It's a letter addressed to you," I said.

"I didn't ask who it's addressed to. I asked what it is." I read on.

"It appears to be an offer for Contractor X to wrap up their work on Study Y that will cost another $500k. It looks like there's a scope and manning plan that justifies the cost as well." I was reading quickly.

"That's what it looks like to me too," he replied. "Book a meeting with Contractor X and get them all out of the building today. I don't want to pay anything more. The study's finished, get them gone." Wow, that was direct.

"They're not my contractor and it's not my study." I pointed out some obvious facts.

"I didn't ask if it was your study."

So, I booked a meeting with the Project Director from Contractor X and we went through their proposal. In the end I suggested that he walk out on to the floor and say in a loud voice, "Everyone working on Study Y: Stand up; put your pens down; and leave the building."

Remaining very professional, he started to slow the cadence of the conversation, saying:

"Well.

"Hmmm.

"We could do that…" and before he could finish his sentence I jumped back in with:

"Great! I'll tell Angus that you'll all be out by the end of the week. Can't cost us any more than 20 grand."

Silence.

The whole industry was overpaid. I could see this clearly because I had come from a regional area where you had to work a lot harder for your significantly lower pay packet. I doubled my salary by moving to Mining. The workforce was mercenary, moving from one project to another continuously. Eighty per cent of the engineers were contractors (rather than employees); they were paid by the hour with a 20% loading to account for the fact that they could be dumped with an hour's notice. Work for 12 months was considered job security. It was a strange market. I didn't feel really good about shutting down Study Y so suddenly. It helped that I didn't know any of the individuals working on it. Most of them would have started on other projects within weeks. I didn't have a relationship with any of them, which I guess is why Angus brought me in to do this job rather than the Project Manager who was managing that study. The power of relationships is real. In fact, it's everything. If relationships are strong, there is loyalty, security and trust. If the relationships are weak or non-existent, then there is nothing.

I started spending time with each of the people who worked for me. I invited the leadership team to my house with their families and cooked a six-course degustation menu (I love to cook) matched with different beers for each course.

(Degustation menus are normally matched with wines. But wines can be a little tricky if you don't have a sufficiently developed palate. In a nutshell, my rule of thumb is that on one end of the wine spectrum exist the two genuine 'big reds:' Shiraz and Cabernet Savion. Both of these are fine drinks. But as you move along the spectrum things get very hazy very quickly. Next to the big reds, is Merlot. This grape makes a wine that is red but not really 'big.' It is challenging to find a situation - or meal – in which a Merlot would be superior to a Shiraz or a Cab Sav. Next along from Merlot is Pino Noir. Pinos are red but even less 'big' than Merlot. In most circumstances a Merlot would be better than a Pino. If you take another step down the wine continuum you get to Rosé. Now as the name suggests Rosé is a pink wine. Its not even red. Then you get to a series of white wines that go from sweet to dry that can be paired with fish and chicken and other white foods. But many beers also pair quite well with white foods. This leaves me to conclude that each end of the spectrum are the places to focus your attention. That is if you're going to step away from the two big red wines, it's a slippery slope all the way to beer. My viticulturist cousin may not agree with everything in this paragraph. But you can match courses to beers quite easily. Especially because no one expects it to work very well. When making invitations to the meal, you send each man away to find a different beer. For example, ask one to bring a six-pack of Spanish Lager (for the antipasto). Ask another for an English Red Ale (for the entrée). And a third for a dark hearty stout (to have with main course or desert for that matter). The men then have the enjoyable experience of wandering into a bottle-o and asking for a Spanish Lager and not really knowing why. The more exotic the beer, the harder it is to find. When the hunt is finally over the satisfaction of having achieved something truly extraordinary cannot be denied the man. When the guests arrive, and beers matching the description are produced at the beginning of the meal, it is usually accompanied with a conversation-starting opening statement along the lines of: "Well, let me tell you, that was NOT easy to find...")

Basically, I started to behave as though we were friends, and magically, we became friends. Some people say that it's harder to manage subordinates if they're your friends. I think that is a cop out – settling for mediocrity, rather than striving for perfection.

It was difficult, later on when the market turned, to give my friends a pay cut, but I managed. I made sure that I didn't ever lie or manipulate them. I just kept to the respectful truth even if it was awkward. Only one person left the team when I handed out wholesale pay cuts of 15%. The industry had turned down severely; the boom was over. The lady who left had her own reasons and really should have left years before. This was the catalyst for her to make the change. Everyone else wanted to stay despite the pay cut. They wanted to stay in the high performing team that we had created, and they wanted to remain working for me. The power of relationships.

(At one point we were running out of money before the next round of study funding was to be approved. The normal approach at this point would be to sack a few people until we had enough money left to keep paying everyone else until the next tranche of cash was released in about three months. We were about twenty percent short of what was required to keep the whole team. So we took the decision to reduce everyone's hours for that period. For the next three months, everyone is having Fridays off. Twelve four-day weeks. Twenty percent fewer hours at the bargain price of twenty percent less pay! They took a bit of convincing, but the alternative was to sack twenty percent of the people. We held a three-hour workshop with the whole team to go through the situation, the decision and to help everyone brainstorm ideas of how to make the most of the next twelve Fridays. One of the draftsmen (or drafties – they guys who 'draft' engineering drawings), Monty, said that there was no way he was telling his wife that he wasn't working on the Fridays to come. He'd just get dressed and leave the house at 7:00am as usual and go to the pub all day. Each to his own, I suppose.)

Then Angus left. He basically woke up one morning and decided that not having a job would be preferable to continuing to work for *the corporation*. Did I mention that he was rich? And I was poorer for no longer having his company. Luckily, I went back to working for Mutsy, so it wasn't all bad.

The corporation was a giant multi-national slow-moving company. We didn't really understand how conservative we were. If conservatism was ranked on a scale of 0 to 100 (0 being low or progressive, and 100 being super-ultra-conservative) we would rank at 92. To the outsider, this would seem to be conservative. But our primary competitor was another multi-national company that was even more conservative. It scored 96 on the scale. This phenomenon warped our view of the world. It meant that instead of perceiving the full 100-point scale, we sort of thought that 92 was the progressive end and 96 was shockingly slow and cumbersome. Still, I suppose that a company only becomes conservative when it has something to conserve, and boy did we have something to conserve – staggeringly high profits. This conservatism kept us all over-paid.

◊◊◊

A new initiative entered our collective conscience. (We regularly had initiatives poured on us. It is what happens to an organisation on the low end of the 92 to 96 scale). Managers started using it in meetings, monthly reports started measuring it, and my colleagues and I in middle management were given a slide pack to "cascade" to all the employees. (This is a technique where someone very, very important puts together presentation material and writes speakers' notes. The presentation is distributed to all of middle management, collectively referred to as "leaders." The middle managers then present the slide pack to the whole workforce. This gives the workforce the benefit of hearing about the initiative from someone they know. It gives middle management the benefit of feeling important to *the corporation* – mainly due to being addressed as "leaders" in the email that had the slide pack attached.)

The initiative was "diversity and inclusion." The two nouns were always joined together. It seemed that diversity could not live without inclusion. And inclusion had not a home without diversity. I was halfway through cascading to my team, when I read the slide: "Diversity and inclusion is the biggest challenge facing *the corporation* today." One of my guys stopped me at that point and asked what that meant. It made me think: I read out the speaker's notes for the slide and we were none the wiser. Considering that our part of *the corporation* basically constructed new infrastructure, it did seem a strange statement to make. The guy said: "I just think that if diversity and inclusion is 'the biggest' challenge we face, then we're pretty much home and hosed." I sort of agreed, it didn't seem like much of an issue, especially when compared with going over budget, delivering the project late or having injuries.

Soon after, I was asked by Lauren, the Senior Vice President (or VP) of HR for feedback on how the cascading went. (I never came across any other VPs in this *corporation*. Or a President either for that matter.)

I thought that I should be truthful. I said that my guys were good; they appreciated the cascade and were interactive, but we were confused by the statement that diversity and inclusion was the biggest challenge faced by *the Corporation*. I had to admit that I didn't really understand what was meant by that, either. Lauren started to bristle, and I realised in that moment that she must have written that slide, if not the whole presentation. Calmly she asked me: "Does your team know what other teams do in our business? Do they reach out to understand people outside their own group?"

"No," I replied. "They understand each other's jobs pretty well and we do a lot of work building relationships so that they are all actually pretty good friends. That, I think, is at the heart of why my team is such a high performing team. But I don't think that we, or I, for that matter really look outside our team much at all. I don't see the value in doing so."

"But," Lauren countered, "do they wonder about what other people's experiences may be?"

"No," I had no idea what she was talking about.

"So, you see why this is such an important issue for us." This wasn't a question and it certainly wasn't an answer either. I could tell that I was dismissed. I avoided conversations with HR whenever possible.

Later on, I worked it out. We all thought that diversity meant having different points of view represented in the lead-up to decisions. We thought inclusion meant that people or groups were not excluded from decision-making bodies. But what *the corporation* really meant by the magnificent bi-fold catch-all *diversity and inclusion* was achieving a 34% target for women in management positions (*middle-management* and *upper-middle-management* really, while the truly important decisions were still made by the two or three guys right at the top of the pyramid). It was the illusion of diversity only. Eventually, the *diversity and inclusion* initiative was re-badged *gender diversity* and everything made sense again... for a short while.

It was suggested by one of the long-term workers at *the corporation* that there were two kinds of people who fitted in. Firstly, there were people who enjoyed yelling at other people. Secondly, there were people who liked being yelled at (or who could at least tolerate it indefinitely). If you didn't fit into one of these two groups, you were never going to fit in. This was a very simplified model, but it rang true. For the record, I don't fit into either of the 2 groups. Tick. Tock.

◊◊◊

One day, Mutsy came up to my floor (this is worth noting, for although Mutsy was a great guy who I enjoyed working for, he walked on to my floor only twice in about three years. The other time this happened, I had been out of the office, but my secretary rang me to share the good news of his presence). We went into my office and shut the door. (I often shut the door when I was meeting with people, not because I didn't want anyone to hear what was being said, but because the air of secrecy implied that the conversations in my office were of the utmost importance and by association, so was I.) Mutsy said that *the corporation* had just advertised a General Manager role in North America. The job was essentially the same as his job (apart from being in North America) and that I should apply, because I would be good at it. I think he expected me to be a little more enthusiastic than I was. I pointed out that if I were to get a job at 'his level', I would have to spend more time with other people at 'his level.'

"Yeah, so," he replied.

"The more time I spend with your colleagues," I said, "the less time I want to spend with them."

"Why?" he asked.

"Because, with very few exceptions, they are selfish arseholes who are only interested in their own career advancement."

There – I had said it! I didn't like *the corporation* or the ranks of grey men that were attracted to becoming senior managers. There was a 'type.' We at worker bee level would describe certain people by saying things like: 'He fits the mould;' or 'He's management material.' The 'type' was grey (wouldn't be noticed in a crowd), never made a mistake (normally achieved by never making a decision), and compliant. When you asked a senior manager what he thought about a contentious issue, you could see his brain whirring as he asked himself the question: "What would the CEO like me to think about that? Because that will be my position."

(Having witnessed this strange occurrence a few times, I thought that my project leadership team might even be doing the same thing, trying to align themselves to what they thought I would want. If that was the case, it would be a disaster. If all my team only wanted to think like me, then collectively we were no smarter than I was on my own. So, I started setting the scene for discussions differently. I would say something like: "Today, I've pulled you all in to discuss this problem. At the end of the discussion, I will make a decision regarding how we are going to move forward. I expect that you will each bring your own point of view to the table, and if you feel strongly about it, you should argue strongly. When I make my decision, I expect that we will all comply regardless of whether your view wins the day." This approach led to great meetings. No-one was afraid to voice their opinion. And unlike so many of my grey bosses, I wasn't afraid to make a decision. By doing this, my leadership team was able to draw on our diversity and collectively be much better than I could be on my own.)

So, I had just insulted the entire cohort of senior management at *the corporation* including Mutsy, albeit inadvertently.

Mutsy said: "Yes, they are. But don't let the arseholes have all the jobs. You should apply, because you could do the job better than them. You probably won't get it, but you should apply anyway."

Mutsy made sense; he had been around *the corporation* for a while, and he didn't fit in. We all wondered how he got his job since he did not fit 'the mould'. I thought about it for two days and then decided to apply. Despite being eminently capable, having proven myself for years with great results, it was unlikely that I would actually be selected. I was a man and appointing men did not support the current gender-diversity drive, and there would be other General Managers from around the global business who would apply regardless of whether they had any expertise in delivering construction projects. (People who were already GMs were superior to other people who were not GMs. I think this was because the company was run by GMs, and their hold on power would diminish if it was easy to become a GM). I could apply for this job in North America and be unsuccessful, but *the corporation* would know that I had GM aspirations. I could then be considered strategically for other jobs that came up in Australia. I was playing the long game. I would be applying for this job in order to get another job in the future. I had become a master strategist … I thought.

I applied for the job. This involved going to an intra-net site, filling in some forms electronically and attaching a copy of my resumé. Having applied, I thought that I'd do it right, go the whole hog and make sure I was noticed. I rang Lauren, the Senior VP of HR. (Retrospectively burning my bridges with her over the diversity thing may not have been the best move. Although the bridges weren't burned to the ground or anything like that, they were certainly charred darkly.) I made up an excuse that I wasn't sure whether my resumé had attached properly, could she have someone check, so that if there was a problem, I could send it through anyway. I didn't really think that there was a problem with my application – I just wanted her to think about me and make some noise with whoever from her team was handling the recruiting. To stand out from the crowd, infamy is almost as good as fame. To my delight, I was shortlisted.

I was contacted by Melissa, a HR recruiter who worked for Lauren and was based in another city. She set up a phone interview with Mutsy's boss, Felix, who was the decision maker. I found this interesting since Felix was already my manager once removed (MoR was the acronym; it's a lot easier to say than my boss' boss). We had already spoken directly on a number of matters, and he sat in the same building as me. In fact, he sat on the same floor of the same building. His office was the big corner office with the best view and a TV screen hanging from the ceiling. It was situated 20 metres from my own office. I thought that Felix and I could have met in person to talk about my application, and since he already knew me, it would probably be a brief chat. But they chose to set up the interview by phone during a week when Felix was doing an international trip.

I have attended many interviews in my career, but this was the weirdest interview experience by far. The day of the interview came. I got up early and went into work early to be ready for Felix's call at 7:00am (this time was presumably convenient for Felix who was in Denver, Colorado at that moment). My excitement built as the clock approached the appointed time. By 7:05 I was getting nervous. By 7:15 I was checking my emails to check that I hadn't got the wrong day or time. And by 7:30 when Felix had still not called, I was really frustrated. I could have slept in!

I wrote to Melissa (and copied in Felix) to ask if she was aware of a change of plans. She promised to follow up. Imagine if I was an external applicant and was sitting in reception for half an hour past my interview time, to be told that the boss man actually couldn't be located, but it would be followed up. There is no way that I would have continued with the application. These are not the actions of a man that I want to work for, or a *corporation* that I want to be part of. This started to sully my view of the greatest company on earth.

A new phone interview time was set up and Felix exceeded my newly lowered expectations by ringing on time. He very carefully said that there was an issue with the last interview time, and he had been tied up. I remember thinking it strange that he didn't apologise for not ringing. He didn't apologise because he wasn't sorry. I had presumed that he was tied up. I didn't think that he was sitting around doing nothing when his phone calendar buzzed with 'time to call for the phone interview with your subordinate!' So, whatever he was doing at the time was more important than ringing me. It was also more important than sending me a text or an email to say that he needed to postpone.

But despite the imposed difficulties of international travel and time zones, we had finally connected. I thought that he would have some questions for me so that he could weigh up my merits against the other applicants. He didn't have any questions: he spoke for about 45 minutes generally about the role and about how good he was. At the end of his monologue, he told me that he thought I was capable of doing the job (I presume that he formed this opinion from talking to Mutsy, since it couldn't possibly have been solely from my polite listening skills). But there were three applicants, all Australians, and the other two were already GMs in the business. I was up against superior beings with little chance of success. It did feel odd that during the interview he had already started the debrief for unsuccessful candidates. I suppose it saves time in the long run. It would have saved even more time since he'd already made his selection, to have not had the farcical interview at all. I accepted defeat and went on a month's summer holidays, thinking the whole sorry tale was behind me.

When I got back to work after a month of bliss, I found an automated email from *corporation* recruitment. It was carefully worded and somewhat cryptic. It read:

> "Following the conversation with you, this email is
> to confirm that your application for the position has
> been closed. We will keep your resumé on file."

The corporation was so politically correct that it wouldn't even say that I was not chosen. *The corporation* was also so unprofessional that the conversation with HR, referred to in the email, never happened. All in all, I was a bit relieved. I didn't really want to move my family to North America, and I wasn't even really sure that I wanted to be a GM. But I was vexed by the impersonal email and lack of heart. I felt that I had an obligation to give Felix and Melissa the opportunity to see that they could have handled this better. I was a person after all, not just an economic unit.

I forwarded the email to Melissa with the following note:

> Dear Melissa,
> Since the referenced conversation hasn't taken place, I will assume that as well as being closed, my application was unsuccessful.
>
> Regards, Jack.

Melissa rang back immediately and said that the email was sent in error and was unexplainable. No decision had been made. My application was still live. It felt like the circus had come to town.

Three weeks later, Felix rang me and said that he had decided to withdraw this GM position. He didn't want to spend the additional cost of an international transfer (we both knew that *the corporation* had been in a very public cost cutting phase for the last 12 months). Felix had decided to appoint one of the existing Project Managers in North America as an acting GM (this is the way you can get someone to think that they're in the GM gang without paying them like the other GMs. I think that acting GMs also have to give their lunch to the real GMs when they all get together for strategy sessions).

I politely thanked Felix for the call, and he politely thanked me for my interest. The final solution was actually quite good. It fitted the cost cutting priorities of the business, and Felix obviously had at least one guy already in North America who could do the job. But it left me with a niggling thought: if the Plan B was so good, why wasn't it the Plan A in the first place?

I set about making myself redundant. This was the golden year of redundancies. The *mining boom* was over; the gargantuan profits were declining (reducing to mammoth proportions only). And *the corporation* was on a drive to increase efficiency (this is code word for getting rid of people). The usual way a company goes about reducing numbers is as follows (this is hard to believe but it occurs with staggering repetition):

o   A target is set from somewhere above (in the adminisphere) dictating how many people need to be severed from *the corporation*;

o   Stage 1: Find some expensive people (managers) and make them redundant. (This means paying them out something between three months' and nine months' pay in return for not coming in to work. You need to get rid of some expensive people, because otherwise you can't really reduce the operating cost);

o   You see how much it cost you in payouts to get rid of expensive people and faint;

o   Stage 2 is then triggered: Find some cheap people (workers) and make them redundant. (Workers attract a much lower payout because their rate is lower and they get more like three months' pay end of the spectrum.)

o   Having achieved the targets, congratulations all round (apart from those who got axed of course) and the new era begins. It is important to note that the 'new organisation' is left with only the people who were situated in between worker and manager – the people who neither know how to manage nor how to do the work.

We had been told by the CEO that this year was going to be "the hardest year we've had!" I interpreted this as "if you can't get made redundant this year, you'll never do it.

Angus had told me previously that it was a good career plan to target two redundancies. One in the middle of your career (to give you a cash boost) and another right at the end (instead of retiring, you can go out with a wheelbarrow full of cash – and then retire).

There was a GM in *the Corporation* who ran his group like the KGB. He was Serbian after all, so maybe he had firsthand experience. His name was Borislav (or just Boris behind his back – a long way behind his back). He loved shooting and saying no (while smiling). He ran utilities. Power and water supply to all the mines. This is bad for project managers like me because every project you ever do will have an effect on water and power consumption. So, everyone had to interface with Borislav regularly. My first encounter was to knock on his office door. He looked up and invited me in. I explained the project we were working on. He listened until I was finished. He said: "I will give you my man Vaughn. He will come to your meeting. You explain to him what you need and he will help you. If you don't get what you need from Vaughn, you come back and see me, Ok?" If Vaughn was going to help me why would I need to come back to see Borislav? Well, that became apparent later. It seemed that once I left Borislav's office, he probably pushed the secret button and Vaughn stepped through the secret trapdoor in the wall. And Boris give very clear instructions: "You go to the meeting. You listen. But you give them nothing! NOTHING!"

Vaughn came to our meeting. He listened to our plans to build a new power line. At the end he said, "we won't approve that."

"What exactly won't you approve?" I asked.

"The power line to the new mine." Vaughn replied.

"Well," I said, "how about this: We'll build the mine, we'll build the power line, ready for energisation. We'll have the new mine ready to produce ore and then let's see how long you 'don't approve it' for."

I was back in Borislav's office a few days later seeking resolution as he had instructed. He came around his desk put his hand on my shoulder and said: "I will fix this for you. You will be able to build your power line. I will make sure it is approved.

"And I want you to remember this."

Whaaaaat! Here I am, just doing my job and now I owe Boris because I'm building a power line! This was messed up. All my colleagues had similar stories about Borislav.

Then Mutsy got chopped. And just like that, the last guy I liked working for was gone. (He was fine; he got the second redundancy of his career as set out in Angus's career advice. He was ready to retire and even had his own winery cellar door to manage. I said that he was Sicilian.)

◊◊◊

I had some exceptional people that I worked with at that time. One was my Contracts Manager named Collette Cash. It was a fantastic name; she was a contracts manager named Collect Cash. She was destined for a career in contracts, sales or debt collecting. She read all the documents that I had to sign to make sure that they were perfect, and she kept me from signing anything that I wasn't allowed to. This left me with time to spend my days talking to my people. This is really important; I've found that people respond best to being communicated with by old fashioned means – using soundwaves emanating from the mouth that enter the other's ears (and not too loudly). The other options that are commonly used are:

- o email (good for transferring data, bad for gaining trust);
- o text message (good for telling simple jokes and texting in sick, bad for everything else);
- o sending a presentation (only good for showing off how well you can do graphics and animations);
- o video conference call; and
- o shouting (bad in almost every situation).

People are people. We like being talked to personally. We respond to people who take the time to speak to our ears, and then allow us to respond back to their ears. I had to sit through a video conference call once with not one but two of the really big bosses on the split screen. They were sitting in their offices staring at their computer screen explaining to the masses (who had logged in to view the spectacle) something about our strategic direction. Now we all appreciate that having a strategic direction is important to any company, but to be honest, we weren't really fussed what it was. Strategy is good, direction is inevitable (unless you're standing still), but we were really just interested in doing our work. We had all logged in out of a sense of duty to the bosses. And they bored us senseless. I wondered if they had sat down preceding the 'show' and discussed what they wanted to achieve by the conference call. Maybe they had the stated aims of gathering a cross- section of the population from around the world and subjecting them to extreme boredom and seeing how they responded. If that was the aim, then they achieved it with flying colours. If there were any other aims, they failed. I wondered how many of the audience were actually listening. I wondered how many were deleting emails while the 'show' continued on their computer screens.

I saw an episode of South Park once where the characters thought that if they shoved food up their bottoms, eventually by a process of reverse-digestion, they would crap out of their mouths. I couldn't help but think that if either of the speakers were to crap out of their mouths on to their keyboards, would anyone even notice? And if they did, would they try to emulate this new management practice?

How powerful would it have been if the guys on the screen had come to my office (since they both worked in the same city, indeed in the same building, as me) and sat down and told me (with mouth-to-ear words) what they wanted me to achieve. They could have then spent five minutes asking me about me. Like a faithful terrier, I would have been loyal forever after. I didn't feel a lot of loyalty to *the corporation* because the two bosses that I liked (Angus and Mutsy) had gone. The new guys didn't know me, and I didn't know them. And what's more, by their actions, they didn't want to get to know me. I couldn't see any reason to try to get to know them.

As well as Collette (who was often just called Cash, or CoCo), I had a Project Engineer (or PE – PEs are young engineers who are being assessed as to whether they can be trusted to be a PM one day) called Benjamin Brothers (we called him Bruv). I had spent a year preparing Bruv to take over my job as Project Manager. I had stated to one of my Manager Once Removed (or MOR) interviews that my development aim was to develop Bruv so that he could do my entire job. And by the following year's MoR, I would be saying that *the corporation* didn't need me because my guy now does the whole of my job, and I was ready for either a more senior position or the golden handshake of redundancy. It took a year to teach Bruv everything I knew (I was surprised that I knew so much stuff that it took a year to get it all out). This is what I taught him:

- o   You have to be present to influence;
- o   Use spoken words wherever possible to communicate;
- o   Speak like a normal person if you want the guys to listen (not like a politician);
- o   Control what you care about;
- o   Relationships are everything;
- o   Guys who like you will go out of their way to make you successful;
- o   Hold people to their promises;
- o   Be firm but fair with contractors;
- o   If a contract ends in dispute, you've failed as a PM;
- o   Never pass up an invitation to a football game;
- o   And if you're really stuck, ask Collette.

Bruv was a smart guy, and I thought that he could go further than me in *the corporation*, mainly because he was more patient. He could remain calmer for longer. I had told him that he was the heir to my fiefdom, and that I had set about making him the new lord (whether he wanted it or not). Eventually, Bruv took over more and

The Corporation135

more of my activities, until I realised that the magical time was upon us. Bruv was doing my whole job. I was just signing things that needed to go further up the line and going to site for two days a week to enjoy the $3 beers at the wet mess for Cheap-Beers-Tuesday nights.

◊◊◊

The whole team was humming; the human machine that I had built was operating without me having to drive it. I could sit back ... and get incredibly bored. I realised that I had achieved my goal one day when I went home early at about 10am because I had new lawn to put down in my back garden. I took one call from Cash who scolded me for bludging. All in all, the day was a raging success. The new lawn was sewn and the team sorted out their own issues without needing my intervention.

When I returned to work the next day, I found a pizza box on my chair. Amusing, although I wasn't exactly sure why. I found Cash and Bruv, and they told me that in my absence they had placed the pizza box (no longer needed for its primary function since most of the pizza had been consumed) on my chair to be my proxy. To cap it off, Cash told me that the pizza box was a really effective manager and they both felt that in the last day they had been given slightly more attention than usual. I was delighted; I could now delegate my authority to a Pizza Box and nothing would change.

When I was alone with the Pizza Box, I drew a circle on the front and split it into four segments, and wrote a few words in each segment. With a broken pen in the centre, we had a 'spin the answer' pizza box. I showed the guys and said, "If I'm not here, you can always ask my delegate the questions that you would usually ask me. No matter what the question, the Pizza Box can answer as well as I can." The possible answers were:

1.Approved;
2.Tell them to get stuffed;
3.No; and
4.Ask Collette.

And the informal instructions to the practised user was to tilt the box so that option 4 was at the bottom, and the pen would naturally point you to ask Collette. Then you could get a more considered opinion. This worked for everyone, even Collette.

Bruv was stoked; he would begin discussions with: "The Pizza Box told me that this was ok, but I just thought I should run it by you since you're here." Of course, I trusted the Pizza Box. If I didn't, I would never have delegated my authority to it, so I would always support its decisions.

Another inspiring dude I worked with was Jimmy 'the Kid.' The Kid was an entrepreneur trapped in the body of a cost controller. We talked for a couple of years about how he wanted to do more with his life than turn up to *the corporation* and be pacified with lots of cash. He wanted to do something meaningful. This rang true for me too and we talked often. Then one day, out of the blue, the Kid decided to put his money where his mouth was and gave up his six-figure salary in favour of the opportunity to have a more meaningful life. He didn't have a plan, but he knew what he didn't want. He didn't want to spend his life in a 9 to 5 job that took all his time from him, so he couldn't pursue his passions, and at the same time be soothed by over-pay that stopped him from pursuing anything else. At the time of writing, the Kid was last seen in South Africa getting his pilot's licence.

◊◊◊

There is a common belief that a successful multi-national *corporation* would have an internal management and decision-making structure that enabled timely decisions to be made in the best interests of the organisation. One would be forgiven for assuming the following:

o   Decision-making committees exist at all levels of the company;

o   The committees are populated by suitably qualified and experienced experts;

o The expert committees would examine the best information available and make decisions;

o The decisions would be clearly communicated as directions to the lower-level officers of *the corporation*;

o The officers would then conduct the business of the corporation according to these directions; and

o As the business objectives changed over time, new information would be digested by the committee and new directions issued.

However, this doesn't actually happen. The greatest company on earth had a system where, for any project to get approved, it had to be reviewed and endorsed by multiple review groups, whose main job was to slow the project down. The first decision-making committee was populated by a group of General Managers and was named "the General Manager Review" or simply "the GM Review" for those of us in the know. Only people at Manager level presented to GMs in this setting. Those below Manager level were glad that they never had to be publicly grilled by the smartest people in the business. In essence, any GM whose portfolio touched the project was invited. And, if this illustrious group was supportive, then the project must be good. The path through a GM Review could be perilous – all the important GMs asking probing questions about your project. However, there are three important things that you needed to know, to navigate the treacherous road of the GM Review:

1. GMs are generally not that clever. Most of them have been promoted to these roles because they have proven themselves to be very good at kissing arse rather than because of any high degree of actual competence.

2. No-one understands your project as well as you do. In fact, no-one else really understands your project at all. Everyone else only knows what they are told about your project – generally by you. So, by following this logic, you basically become the single source of truth (where your project is concerned). Important people are not going to spend their time sifting through the hundreds of report

pages (that you have complied) to understand the project. They will simply demand a briefing (from you). This means that you can present whatever aspects of the project that you like. Obviously, you can't just blatantly lie, or present someone else's project. But you can emphasise the bits you understand and gloss over the bits that you do not. The GMs will never read enough background to contradict you. Occasionally, you can get a question to which you don't know the answer. Some consider this the worst thing that can happen: being asked about an aspect of your accountabilities and not knowing the answer! Shame! However, this tension is simply and quickly dissipated with a commitment to finding the answer to the question and forwarding it to the group later that day. Whether or not you actually forward the answer is immaterial since you are talking to very important people. They will all have many other equally important matters to attend to that day (and every day) and will have forgotten your project entirely within 30 minutes of leaving the presentation room.

3.Not all GMs are created equal. Of the 20 or so General Managers who are invited to the review session, only about three carry any weight. There may be only two of these influential GMs, sometimes there are four, very rarely five, but never six. I shall refer to this group hereafter as the Three Kings. The Three Kings need to be treated very differently to the other lesser GMs. In the weeks leading up to the GM review, it is wise to meet with each of the Three Kings in person to present the project in a private briefing. It's important to always answer the Kings' questions respectfully, no matter how silly they may be. The purpose of these sessions is to gain support from the only votes that count. Failing to gain the support of each of the Three Kings puts you at significant risk. The other GMs (who may be represented as Knights, Bishops and Rooks) know that they are not Kings, yet will be on the lookout for an opportunity to become one. Of course, becoming a King takes time, but bashing a Pawn mercilessly in a public setting might start the ball rolling. However, even a Knight would never attack a Pawn who was backed by the Three Kings. Hence, do not underestimate the value of the pre-meeting briefings. If the pre-briefings are done properly, each of the Three Kings understands the project well enough to ask a sage question during the GM review (to which he already knows the answers). This allows the Three Kings to reinforce their superiority. It also has the added bonus of silencing the other Pieces, who would never ask a "silly question" because this would show them up as inferior to even the Pieces who remained silent.

If the preparation is done well, the actual GM Review is a breeze. After two hours of presenting technical data, assumptions, budgets and schedules (and of course allowing the Three Kings to parade their superior knowledge through questions), the time is up. The audience goes silent for a minute or two and glance down at their watches. Then, one of the Three Kings excuses himself (as he has another even more important meeting to attend, probably with a VP). This leads to a mass exodus where every GM escapes the room as quickly as possible. The Kings are happy for they still hold power. The other Pieces are happy – a good showing is simply ensuring that they were not embarrassed in front of the Kings.

At this point, the Pawns who presented the project are left in the meeting room … alone. Sometimes it feels like we did well, sometimes it feels like we could have done better, but there is always one constant: there is never a statement from the GMs to say that the project is endorsed. There is only silence. The Pawns carefully consider the silence, then inevitably come to the following course of action:

1. The presentation is distributed to the GM audience, attached to a carefully worded email;
2. The email thanks the GMs for lending the project their time and expertise; and
3. States that unless there are immediate and violent complaints, the Pawns will arrange for the project to move on to the next phase of review.

There are never any responses to this email. The Pawns take this as tacit approval, while the other Pieces reserve the right to later say that they never really supported the project should it turn out to be a failure.

So, with the proper preparation, it is possible to garner the support of senior management without a single piece of written communication from them declaring such support (referred to in the common parlance of project management as "approval by silence.") This should be considered a job well done. After all, meetings should only be held to ratify decisions already made.

A successful GM Review meant that the project could advance to be reviewed by the Group Audit of Projects (known as a GAP Review). This was a completely different review. It wasn't about the egos of the Three Kings or any internal politics any more. GAP was responsible for advising *the corporation's* board of management (the biggest fish in *the corporation*) whether the proposed project complied with all aspects of the project guidelines and whether it should be funded or not.

GAP put together a group of independent industry experts (usually old guys who knew their stuff) to review your project. Sometimes this group could be as big as ten experts for a complex project. For a week, this review team would crawl all over the documentation and interview members of the project team. There was no stroking the egos of these guys as they were paid to find problems in your project. I would advise my team to go with the truth. In these situations, it's probably the best policy. If all else fails, you won't ever find yourself unable to remember who you told which lies to. I would also advise them to answer any questions that the review team asked and not to answer questions that had not been asked. You don't need to fill the silence with words. Just wait for the next question. The wily old reviewers had a habit of asking open ended questions like:

Reviewer: "What is the critical path through your schedule?"

Project Team Member (PTM): "There is one critical path that runs through design, earthworks, machine installation and commissioning."

Then there is silence as the Reviewer looks at the PTM. As the seconds pass, the pressure builds. The PTM becomes uncomfortable with the silence. The Reviewer taps his pencil on his blank notebook page as he stares at the PTM. The silence becomes unbearable. More seconds pass. The PTM is overwhelmed with a sense that he has a secret and can't keep it in. The Reviewer looks on in silence and raises his pencil to his lips. A bead of sweat rolls across the PTMs forehead before dripping onto the desk making a small dark blot on his notebook. The silence gets louder. Time slows down and you can hear the slow ticking of someone's wrist watch in the room. The PTM can't take it any longer and is about to explode. Tick, tick…. And then it happens.

PTM: "There are 17 other near critical paths, all of which are delicate, effectively running through all project activities. In fact, almost everything is on at least one of these near critical paths. We have no schedule contingency; we can't monitor and effectively control this many critical activities. In this respect, it doesn't comply with the study guidelines and is highly likely to be delivered late."

The Reviewer at this point takes the pencil from his mouth and jots down a few notes – he has just found his key finding relating to schedule, and all it took was not asking any questions for 14 seconds.

Reviewer: "Let's move on to budget."

A successful GAP Review meant that the project could advance to be presented to the Appraisal Committee (a sub-set of the Financing Committee), who almost always voted with the GAP Review findings. A successful Appraisal Committee meant that the project could advance to the Financing Committee (a sub-set of the Board of Management), who always accepted the Appraisal Committee's recommendations. If the Financing Committee supported the project, it would then be presented to the Board of Management. The Board, with the knowledge that four expert reviews had already been conducted, could endorse the project with full confidence, without having to even read the briefing paper.

◊◊◊

It was time for my next yearly MoR interview and I was excited. I had achieved my aim as put forward a year earlier. I went in to see Felix (the same Felix who had presided over the worst recruiting experience in the history of recruitment). I spent some time in the lead up to the MoR interview asking myself: 'will I? won't I…am I going to tell the truth?...hmm."

Felix had a standard list of questions he asked. The first question was "What are your career goals?"

"Do you mean at this organisation?" I asked back with a little cheeky smile. (It seemed strange to me that the corporation seemed to assume that the only career of any value was at the corporation.

"Yes, at this organisation." He replied with a not-so-cheeky smile.

"Well…" Will I? won't I? … and then I did it: "I don't think that I have a long career ahead of me at this organisation."

His smile faded, "why not?"

"Well, Felix. People here keep encouraging me to apply for GM roles. Which I have done, the most recent of which you know all about. My feedback is always the same. It goes along the lines of: 'We know that you could do this GM role; but there are already a lot of people at GM level who have applied; and It's a shrinking market; and you can't kiss arse.'"

Without smiling, Felix pursed his lips and slowly nodded as he decided not to minute this candid and thought-provoking response. He moved straight onto the next question on his list:

"What are your career concerns, Jack?" At this point I wondered if he even heard the answer to question one. I decided to try another tack:

"Let me put it this way, I don't know why you still pay me." Maybe that would get his attention.

"Why do you say that?"

"The organisation has given me two very capable, smart, young Project Engineers. I have split my projects amongst them. Between the two of them, they manage my entire portfolio. I just come in to sign the things that the 'Delegations Manual' demands that I sign as they cannot."

There was silence for a few moments following that. Eventually Felix replied: "Does your team know that you're unchallenged?" Interesting question. I went with the truth, the full truth, and nothing but the brutal truth:

"Yes, they do. I find it best if I come in late and leave early because I'm less destructive when I'm not here." What would he say to that? Would it arouse some anger from the practiced corporate politician? Or would he realise that I could be removed from the payroll with no detriment to his portfolio? Felix put his pen down, abandoning any hope of taking notes for HR from this interview.

"Well, you're doing a good job, keep it up." I was surprised by this response. It seemed that no matter how redundant I actually was, I couldn't seem to be officially recognised as such. I realised at this point that the only way I was going to get out of *the corporation* was to resign of my own volition. After all, they could never take that away from me!

I comforted myself with the thought that if I couldn't get a six-month payout to not turn up, I could just turn up for another six months and then I would have accumulated the same amount of money. And I didn't really have to turn up that much anyway.

◊◊◊

I was asked to write a document on how to build a positive project culture. It was called 'Culture by Design,' as opposed to the more common 'culture by accident' that occurs on most projects. *The corporation* was one of those organisations that liked consensus (personally, I prefer it if the boss has a vision and some backbone, but alas – Angus was gone). So, when the document had been written, proof read, re-written and generally made perfect, it was then sent out to all the Managing Directors (MDs – the bosses of the GMs) and other selected favourites for review. There was a general idea that this process would allow the document to become even more perfect. Most of the feedback by the important people was of a general nature, and I was able to say that I had "taken the comments into account," while ignoring most of them. I won't bore you with the whole document but, there was one point that got particular attention.

Right after:

o   [Leaders to provide a vision that is bold and compelling], and
o   [Take the time to align the team members to the vision],

I had written the point:

o   [Expel those who wilfully refuse to align].

(Ninety-eight percent of people will go out of their way to do what we ask them to do, so the trick is to express our expectations clearly. But there are the other 2% of people who, for whatever reason, don't behave like the rest of us. So, this point is really important. Alignment on a project is more important than any one person.) The review from everyone's favourite VP of HR, Lauren, came back with this line scrubbed out and replaced with:

o   [Leaders to work with everyone to ensure that the importance of alignment is understood].

I'm sorry, but that is not the same thing.

I only ever expelled one guy who 'wilfully refused to align' and it was the best thing I ever did. He was Wyatt. Wyatt worked for me for about a month when I took over a project. He was a Project Controls Manager (PCM) which is an important role. There are two types of PCMs. The first type translates data into information and gives it to the Project Manager who can make timely and accurate decisions to manage his project. The second type tries to build an empire. Wyatt was the second type. He argued with pretty much all my directions. And actively tried to increase the cost of his company's fee. He had a reputation for taking off on south-east-

Asian holidays when he was needed to present an estimate. One of my colleagues told me that if Wyatt told him that the sky was blue, he would go outside and check it for himself. Not a glowing endorsement for a PCM. The final straw was when it was discovered that he had charged *the Corporation* for an average of sixty-hour weeks for the previous year. He managed this by charging four different projects around 15 hours a week each. This way no single project manager had visibility of his excess hours. In this way he had concocted a way to pay himself more than an MD.

I firstly told him to dedicate 100% of his time to my project or not work on it at all. My project was the best one going around at the time and he extracted himself from the other projects. Once he was dedicated to my project I got rid of him, which meant that he left *the corporation* for good.

He sat in an office where he was visible to the whole project team. He and everyone else on the team assumed that he was untouchable. I told him that we wouldn't move forward with him on the team the day before he was off on an overseas holiday. The next day when he was gone, I had the office cleared out so that he didn't have a space upon his return. It was brutal but effective. For some weeks before that office was filled, everyone looked into the void where 'Wyatt the Invincible' used to sit. I'm sure they thought, if Wyatt can be sacked then I can too; I had better start contributing. The remaining team was stronger without him. I imagined 'working with him to ensure that the importance of alignment was understood.' We would have still been at it ten years later!

It takes a toll trying to keep up appearances when you know that your future is elsewhere. I was sort of threatened twice in that period. I was recognised for being able to build good teams by involvement with the Culture by Design work. This meant that I went to meetings to talk about project culture with important people who didn't know what it meant. In a nutshell, it means trust. If you can build a team that has its own identity, that people want to be part of, in which the people trust each other and their bosses, then you can do amazing things – like deliver projects on time. (It was a privilege to work for an industry that collectively had such mediocre expectations that if your team could only do what it promised on time, then you could be a superstar). It became apparent during these meetings that *the corporation* really only wanted better safety statistics (fewer recordable injuries), and they thought that a better culture would provide this. This is true, in part. Safety performance is a pre-requisite to project performance. You can't have a successful project if people are getting hurt on it. In fact, projects only come in two categories:

1.**Successful projects** have good performance on budget (low), schedule (early), and safety (no injuries). This is achieved by having a project team that understands the project, trusts each other, and commits to working with each other;

2.**Failed projects** are the ones that have poor performance on budget (high), schedule (late) and safety (injuries).

It is pretty rare to have 'in-between' projects that are delivered early but over budget with lots of injuries. They tend to fit into one of the two groupings above. This is because every project team is either working well together or dysfunctional.

The MD who led the Culture by Design initiative kept talking about Safety Culture by Design. I said that the initiative needed to pursue a performance culture, of which safety is a pre-requisite rather than being an end in itself. *The corporation* had been talking up *safety culture* for about a decade. We have been using the following key phrase:

"We want everyone to finish every shift in the same state as they started it." You know, with the same number of arms, etc.

This message was lost on the trades. They always switched off because something was not quite right; something was a little disingenuous. I worked it out. I translated the key phrase above into trades talk:

"We want you all to not deteriorate during the period for which we have some responsibility for you (because if there are injuries, our bonuses will be less)". Not exactly an inspiring and compelling message.

I took a different tack. I told everyone on my project that I expected that they would develop and grow while they were on the project. Of course, I expected that they would all remain safe, but more than that, I expected that when they left the project, they would be better than they were when they joined it. The improvement could be personal or professional, I didn't mind, but they all had to achieve something over and above getting paid for digging holes. It was a raging success – we had guys who gave up smoking, we had guys who got their finances in order, we had guys who got promoted. But within this environment of self-improvement, no-one got hurt. (Well, we had one Kiwi guy who crazily managed to crash a 4WD forklift into the flat piece of ground and then broke the windscreen with his head because he wasn't wearing a seatbelt. He removed himself from site the same afternoon, never to return. But that was an anomaly).

Eventually, during a meeting I told the MD and others that I didn't think it mattered what we did on this 'Culture by Design' program – it wasn't going to have any effect on the way the global organisation delivered projects because it was only interested in statistics not people. The MD replied with:

"Do you aspire to be a GM in this business?" And there it was! Was he saying: 'All that diversity crap we talk about, forget it. If you can't fit in and pretend that this initiative is very important and will change the world, then you'll never get a promotion in this company'? Is there any other way to take it?

"No, I've met too many," I replied. How could he get me so wrong? I don't respond to threats. And what's more, everyone around me knows it. The MD who was charged with changing the culture of *the corporation* to one centred on building trusting teams couldn't be bothered getting to know me. Ironic.

The second threat was during an early meeting with my new GM Ken Delaney, who I called Kenny D (he was one of the greyest managers ever). He said: "You going around telling HR and everyone that you want a redundancy isn't doing you any favours.

They're talking about you at the level above me and I'm just saying that sooner or later someone will see *the corporation* on your resumé and ring someone here. You have always been ranked a number 1, but recently you've slipped to a 2."

The number 1 reference might be confusing. During performance reviews, *the corporation* gave everyone a number between 1 and 9. A talent ranking. A subjective assessment of how far up you can get promoted.

- o   9, 8 and 7 all mean you're crap and probably should be gotten rid of if the opportunity arises;
- o   6, 5, 4 and 3 have different subtleties but basically, you're in the highest job that you'll ever do;
- o   2 translates to one promotion within the next five years; and
- o   1 means capable of two promotions within seven years.

Although you weren't supposed to know what your number was, I knew that I was a 1 because Angus and Mutsy had both told me. And what good did being a 1 do for me? I spent five years in the same job. Not promoted once – let alone twice! Maybe being back in the number 2 spot might have a beneficial effect on my career. But I didn't think that Kenny D would understand my point of view.

What did he mean that it was bad that the hierarchy was talking about me wanting to leave? After all, I thought that the only thing worse than being talked about was not being talked about. When he said that it wasn't doing me any favours, it seemed to me that it actually was working along the way I wanted it to. Was he saying that I shouldn't talk about getting made redundant because … it might happen? (And for the record, I never told anyone from HR about this or anything else for that matter. Everyone knows that HR are not to be trusted.)

The weirdest part of this conversation was Kenny D's belief that those currently employed in senior positions at *the corporation* had power over my future career (even outside *the corporation*) and indeed my future happiness. The arrogance of the organisation I found astounding. But you had to remember that this was the greatest company on earth. It was convinced that the rest of the world was inferior….and may not even exist.

You know, I think that Kenny D actually believed what he was saying, and he wasn't really making the threat that I perceived. I think that he just believed the bullshit and was loyally passing on the messaging from on high. The best bosses I had wouldn't have done that.

At this time, the most touching thing happened to me. I found on my computer keyboard a package about 10cm square with a printed note on top that read, '*I think you're awesome*'. It also had a number 1 in a shield above the text. I unwrapped the package to find a caramel slice. I was hungry and the caramel slice was delicious. But the note almost made me cry. For someone to think strongly enough to go to those lengths to anonymously pass on that message meant a lot. All the time I had spent caring for my people was noticed. I never found out who left the caramel slice, there were a few suspects. Kenny D was not among them.

Bruv and I took a site trip to pretend to visit one of his previous project sites (which we did drive past), but the purpose of the trip was to have lunch at the Point Sampson fish and chip shop - arguably the greatest fish and chips on the planet with ocean views from the northern coast of Australia. We went all out with the meal and had the big seafood basket – including a bucket of Calamari, battered Spanish Mackerel fillets and enough chips to feed a small African nation. It cost about $80 for lunch for two. Expensive? Yes. Extravagant? If you add in the airfares and hire car… absolutely!

Kenny D told some of us PMs that we shouldn't have relationships with workers. He didn't use those words, and he wasn't talking about romance. He had the view that a PM should be aloof; separated from the project team members. Revered from a distance. And when he spoke, everyone should stand to attention in the magnificence of his presence. I was astounded – where did a welder from Northern England develop such a toffy attitude? I guess that someone who is unable to build relationships wouldn't understand the value of relationships. This was the beginning of the end. He clearly wasn't about getting the best out of his reports – getting us to do our thing better than before. He was about getting us to do his thing, his way. I had to get out.

I had a remarkable performance review with Kenny D. I'll recap the standard rule of performance reviews from an earlier section: it doesn't matter what you do, you will score about 75% because lower scores reflect badly on the manager and higher scores imply that you should be the manager. We spent two long hours trawling through my boring answers to boring questions about how I had lived *the corporation's* values, how I had developed talent for the future, and how my projects had performed. At the end, he said: "You have achieved everything that we said was important…" That sounded positive. "…but what you don't do is have an influence in the organisation outside your accountabilities." I thought about this for a moment.

"Isn't that your job as a *General Manager*?" The question was left hanging. "Am I supposed to?" I continued.

"Well, yes. Even though it isn't in your job description."

"I'm confused. If there are other things that you'd like me to do that aren't in my job description, maybe when we're setting performance objectives for next year, you could articulate your expectations and then I'd probably be able to achieve them."

"I'm just saying that when I have problems, you're not the guy who I go to, to fix them." And that is exactly who I wanted to be: not the guy Kenny D leans on for help. It looks like I really did tick all the boxes that year. It occurred to me that Kenny's complaints about me were really comments about his own leadership rather than me in particular. He drove compliance and a philosophy of make-sure-you-tick-all-the-boxes-whether-they're-important-or-not, combined with a regime of if-you-stick-your-head-up-you'll-be-punished-by-having-to-do-more-work. Although I really disliked his approach, I am influenced by my environment, and, like all my colleagues, I had started walking the path of least resistance.

Previously, I had been used extensively by Angus and Mutsy in doing all sorts of special projects that were outside my project accountabilities. Kenny D knew what I was capable of – and he had failed to engage with me, build a relationship with me, motivate or inspire me, or win my heart or mind. As a bare minimum, he had even failed to delegate any tasks to me.

"If you're planning on leaving," Kenny continued, "maybe you should really try to impress us in the next six months or so." It occurred to me that he either lived in a parallel universe in which there were no people (just objects), or, that he was totally out of his depth and couldn't even discern which bits of the HR handbook to use for different situations.

◊◊◊

Kenny D wanted all six of his Project Managers to do a driving tour of all our construction sites with him. He explained that it was important to get PMs to see each other's project sites. The plan was to spend three days driving around in two cars on a lightning tour, spending an average of two hours on each site. The proposal for the three days included a total of 11 hours on sites and 18 hours of driving around the desert between sites. Seven men, two cars, six sites, three days together in close confines: it sounded awful. One of the PMs asked the obvious question: "What are we trying to achieve by this trip, Kenny?"

"For everyone here to be familiar with all the sites."

"But if we only have two hours per site, then no-one is getting out of the cars, since it'll take two hours just to drive around most of the sites," the PM continued.

"I don't think that you're appreciating the value of this trip." True. The value of this trip had neither been successfully articulated by Kenny D nor appreciated by any of the victims. I thought that by telling a story with a powerful imagery, I could break the deadlock in our favour. I said:

"A couple of years ago, I chartered a small yacht for a week in the Whitsundays with six friends and one old dude who actually knew how to sail." (It was a great trip and I highly recommend it, just try not to hit the bottom like we did). "Anyway, the first day when we got the boat, we were all choosing beds and I had an epiphany. I said to the other guys: 'In a week, when we're leaving, this boat is gonna stink.' And it did."

Kenny D and the PMs just stared at me blankly, obviously caught up with the literal meaning of my story. I continued: "What I'm trying to say is: I only mildly dislike you guys now, but after three days in each other's pockets, we are going to hate each other."

The trip was cancelled. Kenny D was not impressed.

◊◊◊

We were an electrical engineer short (imagine almost having a royal flush in poker but missing the jack). So I set about recruiting one. You'd be forgiven for thinking that advertising would be the best way to recruit an engineer. Advertising is only the third best way to recruit an engineer (and highly likely to be only slightly better than just putting up with not having one). You see when you advertise, read resumés, conduct interviews and make a selection of candidates you still basically have no idea who you're going to get. You might get a good indication whether the lucky candidate can fit into your organisation a day or two after he's started. More to the point, it will probably be obvious after a day or two if he really doesn't fit in. However, it may take weeks or months to verify if he does fit in. After all, the ability to do calculations is normally the least important skill to be assessed when selecting engineers. The most important question to be answered is "can the new guy fit into the team and communicate with stakeholders without causing a higher level of trouble than the problems that he solves (if not, this would be a net loss to the employer and is more common that you might think). When Angus got rid of his incompetent project managers he said: "I'd rather have no one than have you. My life would be easier if I personally did your job as well as my own instead of managing you." Harsh. But if it's true, it's true.

(When my daughter was applying for university, she explained to me that she had spoken to the student services people at the uni. She had been advised that the best thing for her to do was to study a double degree, even though this would take five years instead of three. She explained to me that this would mean that in five years when she was applying for graduate jobs, graduates with multiple degrees are more impressive than those with only a single degree (as well as being two years late entering the jobs market). I said: "Hands up everyone in this room who has employed graduates before." There were only the two of us in the room. I raised my hand. "Oh, its just me. And the last thing I'm looking for in a graduate is someone with multiple degrees, or a masters, a PhD or any other form of further education. I'm looking for someone with decent marks that can do what he's told and is prepared to learn how to do the job. All I heard is that the university wants you to pay them for five years instead of three.")

Anyway, the best way to recruit an engineer is to recruit someone who you have worked with before if you know that he can do the job and fit into the organisation (this can be very coarsely referred to as 'employ a friend'). The second-best way is to recruit the friend of a friend. This normally goes down by asking everyone already in the team if they know anyone who is competent and available who might like to work with us. And that is exactly what I did. I was given the name of an electrical engineer from one of my guys who I trusted (someone who is trusted by someone who I trust is already far superior to advertising). And that name was Nick. It was probably really Niκo since he was Greek, but he introduced himself as Nick.

I met with Nick at a café to discuss his potential move onto my project. He wore a suit with an open neck (no tie). This is important. I'm not recruiting a lawyer. If I were, I may have been impressed with a suit and tie. However, if an engineer is wearing a suit and tie, he is almost certainly a pratt. (A mate of mine who started his first graduate job - all those years ago - wore a suit and went straight onto a site visit on that infamous first day of his. To minimise the scorn, he took the initiative with the operators in the field by introducing himself as a "bloody engineer from head office.") Nick was wearing a suit without a tie. This is not ideal but acceptable for engineers. Without the tie his odds are much better, with only a 43% chance of being a pratt. (The standardised engineering uniform protocols recommended clothing that changes during the days of the week. Monday to Wednesday: Chino pants and an open neck shirt. Thursday: Chinos and a Polo shirt. Friday: Polo and Jeans. The uniform regime is carefully designed to ease you into the weekend. Switching up to an open neck suit is only a small increase on Monday, but quite a leap on Friday. I can't remember which day I met with Nick.)

I had a long black with a dash of cream. (This is a very sophisticated coffee – once you've had cream in coffee, you never willingly go back to milk. The cream adds 'creaminess' without diluting the 'coffee-ness'. Diluting the coffee is equivalent to introducing weakness. Not ideal.) Nick had a cappuccino (which is really a children's drink).

Because of the recommendation I had been given, I knew that Nick could do the job, so I asked about his current role and his availability. He said that he'd been in his current job for a few years and the project wasn't getting approved any time soon, so it's a bit of a dead-end. Consequently, for the right opportunity, he was prepared to move on. Solid answer. Despite his suit and choice of coffee, he was a very low risk recruitment and I asked when he could start. He replied: "shouldn't you tell me something about the project first?"

"Good point, Nick. I forgot that I'm supposed to impress you too. Let me tell you about the project. It's a middle-sized project. With the basic elements that are common to most open cut mine developments. Roads, buildings, primary crusher, overland conveyor, tied into an existing processing hub." As I said this, I realised that I was boring myself just describing the project. "But what's more pertinent is that the *corporation* is pretty frustrating to work for. Its more focussed on compliance with largely irrelevant systems rather than achieving results. The people are pretty good as long as you don't meet those further up the tree.

"Let me put it this way, Nick, when you're an old man and your grandchildren sit on your lap an say – '*Grandaddy, what did you do when you used to work?*' you won't tell them about this project.

"So, are you in, or are you out?"

Nick looked at me with a quizzical smile on his face. "You know what, I'm in." He started with us four weeks later.

On day one, Nick asked me what I wanted him to focus on. I replied that I just wanted him to fix the electrical stuff on the project. I don't know what that means, because if I did, I would be doing it myself instead of employing him. He was taken aback for a moment but his immediate surprise quickly changed to relief. His smile said: 'no one has ever treated me like an adult at work before!' Nick wore his open neck suit Monday to Thursday and jeans with a shirt on Fridays. He obviously followed a dress code that had been published some decades before ours. This meant that on Thursdays he was really showing the rest of us up.

On about day thirty, Nick told me that it was pretty frustrating to work for the *corporation*. I said: "You got full disclosure of that over your cappuccino. And you're not going anywhere!" He didn't disagree and he didn't leave.

◊◊◊

Then an amazing thing happened: I was asked to be part of a GAP Review team. Kenny D came into my office and asked me if I was busy these days. This is always a difficult question to answer. You normally don't really want to say that you're not busy because that means that they can give you more work to do. On the other hand, you don't want to say that you are really busy in case there's a junket on offer. I played a straight bat saying: "Well, Kenny, there's a bit on, but what do you have? I guess I can squeeze it in if it's important."

He said that GAP was asking for a project management reviewer, and everyone else was busy, so he'd put my name forward if I wanted. I tried not to let my excitement show.

I had presented to GAP some eighteen times in seven years, so I pretty much knew the ropes even though I'd never been on the review team before. And I knew how to be silent between questions to drag the truth out of project teams. The GAP leads were happy with me and one review led to another. This basically meant that in addition to my day job, I started travelling around the country to do week-long reviews on other people's projects. It was fantastic fun. A review team consisted of a number of technical experts and me (as a project manager, I consider that my specialty is not having a specialty). My role was to review the project management aspects, which I found easy. It soon became obvious that project management to most people was a 'black art'; an alchemy that could neither be understood nor quantified. The biggest difficulty I had was needing to go on a detox the week after a review due to the copious quantity of beer that tended to be consumed in the evenings of the review. Let's just say that it is important to get put in a hotel near the office so that you can set your alarm clock for 8:45am and still make a 9:00am start. Even after I resigned, during my four-week notice period, I reviewed another two projects! (and then went on a thorough detox.)

You read that right, I resigned from the greatest *company* on earth. I started telling people that I meant to move on and the word got out. In a three-month period, I had six unsolicited approaches from friends, acquaintances and ex-colleagues. Five were more of the same and a bit ho-hum. But one was completely different. The opportunity presented itself to join the management team of a small group of companies with an existing business base, and lots of potential, working with the smartest guy I had ever met. Mike had worked for me, on my team, for about two years. He was brilliant. He remains the only person that I've ever met who could do high level strategic thinking one moment and delve into the minute details of a project estimate the next. And he was a people-focussed leader. The young guys talked about him as Magic Mike for obvious reasons. I was lucky enough to have him on my team because at the time, the market was turning down and he knew that he should move out of his corporate role (as an overhead) and on to a project (fully reimbursed). After about two years on my team, my management kicked him out of the business for being too clever. (There were other reasons cited, but the real cause of his removal was because he made the bosses look dumb, and probably made them feel dumb, too.) I learned so much working with Magic Mike, and I had always trusted him.

So, Magic Mike rang me up and offered me a gig working with him in a new little company, and I just said: "Any opportunity to work with you again, Mike, I'm in. You just sort out a contract and pay and everything; make me the best offer that you can and I'll accept it. The deal was that I would work with Magic Mike and also be available to consult externally, so that I could keep doing the project reviews. It was a match made in heaven.

Of course, I had to resign and give Kenny D my notice. My natural inclination is absolute honesty. So, what I wanted to say was: 'Kenny, you're the worst boss I've ever had. You spend your time micro-managing unimportant details. You are an abysmal choice for a GM considering that you have neither a strategic or relational bone in your body. All in all, I consider that I'd be better off having no income than continuing to have to talk to you.'

What I really didn't want to say was: 'It's not you, it's me. I just think that we'd be better off if we spent some time apart for a while. I still *love* you, I just don't know if I'm *in love* with you. Some space between us would give us the chance to understand our feelings better...'

The words I used: "I've had a really big week, I've been offered a job in the management of a small group of companies that looks really exciting on a number of fronts, and I'm going to give you notice today. Considering that, what do you need me to do in the next four weeks to make the transition as easy as possible?" In fact, I had already handed everything over to Bruv about a year earlier, so there was nothing to do.

Notice period is a magical time. Best described as the month when you have not one but two jobs and have to attend to neither of them.

Having resigned, I dished out my trinkets to various team members. Bruv got the Pizza Box and some books. My conveyor scale model went to another project manager. My 'WTF' stamp went to my engineering manager with instructions to stamp it on as many drawings as possible.

Everyone leaves a legacy; the only question is: will it be positive or negative?

There were three leaving events held in my honour: two morning teas and one drinks at the pub, all organised by others. The first morning tea was a bring a plate affair, and only a small sub-set of colleagues were invited. There were about 40 people (half of whom I didn't know), two plates of biscuits, and a whole lot of awkwardness.

The Primary Crusher (who was now an MD) came up to me and said: "Jack, I've been meaning to catch up with you." This was a lie. With the considerable resources at his disposal, if he had wanted to catch up with me, he would have made it happen (he walked past my office at least twice a day to make his cup of tea). "I just wanted to say thank you for what you've done for us here." Oh, you are good! I thought that I'd give this new technique a go:

"I also want to say thanks – thanks for your patience and attention. Although we haven't had the best relationship over the years," this was quite an understatement; I once told him that he treated me as a 'scapegoat in waiting' and refused to work on his project, "over the last six or twelve months I feel that we've been getting along really well and have achieved a lot together in that time." The real reason that we'd been getting along well was because we had had no contact in that time.

We shook hands and smiled, both delighted with each other's kind words, in full knowledge that neither of us believed the other. This was so much more harmonious than being totally honest. Magic Mike had advised me to leave well and not burn any bridges on my way out.

The next day was my actual last day. I spent the day going around the building and saying goodbye. I thought that it was important to do this personally. After all, I had been a pretty visible and outspoken leader over the years. I typed up an all-staff email that concluded with a statesmanlike "… I want to thank you for both the opportunities and support afforded me, both of which have been substantial. Until our paths cross next, thank you and farewell."

Kenny D came to my office at two o'clock in the afternoon to retrieve my company laptop computer. I resisted the urge to point out that if there was anything valuable in it, I wouldn't have waited until my last afternoon to take it (there wasn't anything valuable in it). I gave him my computer and for the next two hours couldn't even pretend to work since I was computer-less. At four pm, I handed in my building pass, parking key, and credit card (cut in half) before leaving the building.

Kenny D put on drinks at a local pub with a small group of colleagues (ones that I actually knew). I thought that I should try to speak to Kenny D in a positive light, and try to make our last conversation (ever) an uplifting one (just like the Primary Crusher had shown me a day earlier). I racked my brain to try to think of an endearing feature of his. It took a while, but in the end, I went with complimenting him on his decency and consistency (he was both of these things). He responded by listing my faults. In summary, and heavily paraphrased, I was a young and arrogant upstart, who wouldn't do what he was told. Essentially, I had no quarrel with this summation. He said that with enough time, three to five years, he could have turned me into a GM. I thanked him for drinks and pondered this last statement. I think that even with 100 years, he couldn't have made me bland enough to be a GM of his ilk.

The next day I came back into my old workplace to attend a morning tea put on by my own team. I had to sign in at reception as a visitor and get someone to come down and escort me into the building. It was a little embarrassing. There was lots of food at this morning tea, and there were a couple of complimentary speeches before I had the opportunity to reply:

"I was signing in at reception 20 minutes ago, and it reminded me of a time seven years ago, when I signed in here while I was waiting to be interviewed by Mutsy. That interview culminated in Mutsy telling me about his experience of driving around Naples and me telling Mutsy about my experience of being driven around Naples. At that point I thought I'd nailed it! Four months later, I got offered this job.

"During one of our secret PM meetings, Mutsy once said: "If you guys think that you're leaders, have a look over your shoulder every now and again to see if there's anyone there. 'Cause if there ain't, you aren't!'" I want to thank you guys for making me the leader that I am. For years when I've looked over my shoulder, you've been there. And let's be honest, the One Million Man Hours worked with only a single injury, the 34% under budget, and the ten weeks ahead of schedule (which are all front and centre on my resumé), weren't really my doing – it was due to you guys. All I did was sack the bad people – that was my contribution. Thanks for making me successful.

"Those of you who were on this team at my start here, seven years ago, would have noticed a bit of a change in me in that time. I was a bit angrier when I started and some of you guys were pretty patient with me for a long time until I learnt to trust you. I was all about accountability and relationships. In that order. I think that I'm now more about relationships and accountability. I used to describe my role as achieving results through people, but now I think that what I really want is to achieve people through results. You guys are the fruit of my labours.

"When I was about 13 or 14 years old, I had just become aware of concepts like compound interest and the stock market. I remember asking my Old Man, "Why don't we have any investments?" He replied: "I do. I have five children." Up until that moment I knew that my dad was clever, but that was the dumbest thing that he had ever said. It's only now, decades later, that I think I'm starting to understand what he was saying. But I want to encourage you guys to continue to invest in each other. And not necessarily for your own direct benefit, but for each other's benefit. And most of all, like any grandmother and Mother Teresa would tell you, look after your kids; it's more important than anything else.

"I wish all of you the very best in the future. If I can help you out in any way, just sing out — my phone number isn't changing. And use me as a referee if you ever need to. Tomorrow or in ten years' time if need be. I'm a big believer in every one of you. You are all standing here because you've been hand-picked and have outperformed every other team in this business.

"Thanks for being my team."

And then my secretary started crying.

SALES

My new position title was: General Manager – Key Accounts. This is code for 'whatever it takes to get and keep a customer'. This new gig was with a very small company, so we didn't spend our time serving systems. We only served customers (and prospective customers). This meant that I gained about six hours a day (at *the corporation* I had previously spent this time doing low level administration).

I agreed to a permanent part-time position that didn't pay much, but I only needed to attend for about 20 hours a week. This was my dream of working part time and having more time for not work stuff.

The office was not in the CBD. We worked out of an old house in a central suburb about 15 minutes' walk from the city centre. It was so much more pleasant than working in town. There is actually more sunshine in the suburbs (because the buildings are lower, the sun can get through). I shared an office, and we weirdly went outside on to the second storey balcony to take private phone calls. My part of this enterprise was:

1. Know what was going on in the industry;
2. Identify needs that matched our products and services; and
3. Make the right introductions at the right time.

I realised that this whole role could be summarised as "I know a guy." This was my usual response when we were doing planning. And I realised that I knew lots of people. Whatever the question, a grand answer was: "I know a guy."

Like: "We need to identify a market for this new product within company X."

"I know a guy."

Or: "We need to find a contractor that we can form a JV with, in order to bid on tender Y."

"Yep, no problem, I know a guy."

We achieved a lot in a short time. And in the background, I was still doing the odd GAP review for the greatest company on earth (as an external independent expert – the word expert still makes me giggle).

(The delight of my first engagement as an independent consultant is forever etched into my memory. My wife called out from downstairs that Yasmina was calling. There was only one Yasmina I had ever met so I knew exactly who she was talking about. Yasmina was one of the GAP leads who I had previously done some work for. So, I knew immediately that this was probably an 'opportunity.' I took two stairs at a time to descend quickly while still being careful to protect my ankles from rolling. I answered the phone while trying to control my heavy breathing from the exertion.

"Hello, Yasmina."

"Good morning, Jack." Yasmina was Persian by descent, but was raised in England. I.e. she looked middle-eastern, but sounded very English. And not a Cockny or Northern type English accent. An English accent that is best described as 'educated.' "Would you be available to attend a review with me in Germany in six weeks?"

"Let me check my calendar," I didn't want to give the impression that I wasn't busy, even though I wasn't busy. After pausing for five seconds or so, just long enough to give the impression that I had a calendar to check, I continued: "Yes, I'm available that week."

Six weeks later I was in Leipzig with Yasmina and the review team. We finished a day early. So, the extra day before we left was spent assisting Yasmina buy shoes in a European department store. I think that this cemented my spot on the team. Not sure if anyone remembers my project management comments, but, but my shoe ferrying to and from the sampling seat was first class.)

After about three months, it became apparent that the little company I was employed by was totally dysfunctional. It was, in fact, a partnership between two South Africans. South Africans come in three categories:

1. The English-type who are pretty normal – they just sound funny when they talk;

2. The Afrikaans-type who have a tendency to be obnoxious; and

3. The women, who are lovely (probably because they have to put up with the Afrikaans men).

One of the partners was English and one was Afrikaans. They pretty much hated each other. The English one was a nice guy but had no backbone. He would always say yes and very rarely deliver. The Afrikaans one was 'a difficult character' who cared nothing for anyone but himself.

(The Afrikaans are characterised as being big and abrupt. I guess that's what you get when you send the Dutch to live in the African desert with only beef to eat. During a rugby world cup, the Springboks (the South African national rugby team) were aptly introduced by a commentator: "For this world cup, as usual South Africa have prepared a team consisting of a pack of enormous, straight-running, hard-hitting forwards supported by a backline of enormous, straight-running, hard-hitting backs." A South African scheduler worked for me over a couple of years, Hendrick. Just like the Springboks, Hendrick was very big and very tough. He lived on a farm a couple of hours from head office and drove into town on Monday morning for the week. One Monday he was late to work and when he came in, he was limping. I asked him if he was ok. He responded:

"Yesterday on my property, I was in bare feet and I jumped over a creek. When I landed on the other side, I accidently impaled myself on a tree branch. The branch went through my calf and came out the other side. It hurt a lot, and there was so much blood. O my gosh."

"Are you ok now?" I was concerned.

"I am fine. I have bandaged it up, it'll be ok. But you know what the worst thing is. I drove to the pharmacy. And you know what, they won't sell you morphine in this country!"

"Well Hendrick, Australia is a bit of a 'Nanny State.'")

I had been working in this place for almost six months when I found that I had not been paid for the previous month. This was strange because although they didn't pay much, they did pay monthly. I went to see the English one to ask why I hadn't been paid for the month just gone, and he said that I wasn't going to be paid, because they were splitting the business or something, and that we had talked about it.

Although we had discussed splitting up the business, the bit where I didn't get paid had never come up. I said: "You do realise how illegal this is, don't you? You can't just not pay me. You need to give me notice under the conditions of my contract. This is Australia."

Things got pretty heated at this point, and he told me that I should get a lawyer. I said that this didn't need a lawyer – an ombudsman would sort it out pretty quickly. He said that I was unreasonable. I asked how he could be so incompetent. He could have handled this situation so differently by being up front and honest with me three months earlier, when he and the Afrikaans one had agreed with each other to stop paying me. It took me back to something I read in my self-driven leadership training:

Even when you're sacking someone, you're sacking a person.

In my time managing major projects, I've sacked some people. I think I got better at it over the years. To do it well, it's actually really simple: give a stuff about the person who is losing their job. The English one didn't care about me. In hindsight, I think that I agreed to a rate that was too cheap, which led to him undervaluing my contribution. I was angry for a long time.

Suffice to say, I stopped working for them at that point (not being paid is a good motivator to stop going to work). Over the next three months, I spoke to the English one three times, who agreed to pay me what was owed, but never actually did. This situation was very difficult: I was in the right both legally and morally. It churned me up inside that this ridiculous situation could come about. How could these people be so inept that they could act so illegally? How could they be so bent that they could act so immorally?

I considered my options and legal training (the word "training" used here is an embellishment – my legal training consisted of instructing lawyers and getting them to write contracts and claims; however, I had spent a lot of time with lawyers, and I knew some of the words that they used) and wrote a letter of demand. I chalked up a large damages bill (as a starting point for negotiating a settlement) and presented it in a cutting but easy to read manner. I even got it checked by a claims lawyer who I had worked with previously. He made it even more compelling, adding words like 'deceptive and misleading' to the text.

At the same time, my work with GAP took off and I was doing about one a month. In fact, I was having trouble finding the time to finalise my letter of demand. In the end, I never made that claim, and they have never paid me for my last month and legal notice period. It just wasn't worth it. Making that claim (for which I was legally entitled) would have caused lots of heartache and stress. For me, for the English one, and probably for the Afrikaans one as well. And for what benefit? Two months' worth of salary. When I was already earning much more than this by contracting. As a compromise, I never returned my work computer.

This whole debacle turned out to be a blessing in disguise. I can't imagine that I would have plucked up the courage to leave the greatest company on earth if I didn't have something resembling security to go to. I would not have dropped the safety net of being a big business employee to go out on my own. In reality, the security of my contract with the South Africans was merely a perception. But it did give me the ability to make the jump, even if it was on to a quickly deflating pillow.

Retrospectively it was like stepping off an enormous ship (the greatest company on earth) onto a small boat (the South Africans' company). The small boat rocked around on the waves much harder than the enormous ship ever did. In fact, it rocked around so hard that it wasn't obvious that it was sinking. By the time it did become obvious that it was sinking, it was too late to do much about it. As the small boat sank and as I began treading water, a life raft (independent consulting) rose from beneath me. Astonishingly I was floating again.

INDEPENDENT CONSULTING

I didn't really have to do anything at this point to become a full-time independent consultant because I was already doing it. The phone was ringing and people (who I already knew) were calling me to ask for my assistance. People from *the Corporation* would ask me how it was being "on the outside". My usual response was that it was great as long as the phone kept ringing. But I imagine that it would get pretty old, pretty fast if the phone stopped ringing.

Usually, when I am working in a client office I am given an access card for the duration of my visit. You need to swipe the access card to gain entry to elevators or doors. It is common for these cards to be issued with lanyards so that you can wear the card around your neck. If this seems a little strange that's because it is. Wearing a *corporation* card around the neck is a bit symbolic. So, it's not surprising that *the corporation* would demand this gesture of its employees. One building required that the card was swiped on a pillar that would open little gates in the building foyer, allowing access to the elevators. The pillar was only about 4 feet tall which meant that everyone had to lower their necks in order to make contact with the sensor. I found it interesting to watch a steady stream of employees and guests, wearing *the corporation* card as a necklace, bowing at the entrance of the building in order to gain entry to *the corporation* temple. I keep my temporary access cards in my pocket.

Another way of describing my new job was a term coined by one of my fellow reviewers in an old German-style hotel in Namibia. There was a grandiose, plush, carpeted staircase that led up to the guest quarters. The front bar was a wood-panelled, darkened room. The fireplace was burning very low (because it was summer and totally unnecessary in the African desert, but a fireplace like that deserves a flame). It smelled of smoke but it reeked of tradition. We were around the bar, sampling the local beers, when the conversation turned sharply to a comparison of the best looking mine pits in the world. The Geos[1] on the review team started putting forward various pits as the "best pit," with reasons as to why it should be considered superior. The deepest, the longest, the largest volume of hole, the most employees, the best safety record, often culminating with the expression: "What a pit!" Quite frankly it was pathetic and I had to ask them: "Are you guys so bloody boring that the only thing left in your lives is perving on mine pits?"

There was a moment of silence. "Jack, it's more like mining tourism."

And there you have it! My new job was mining tourism. I travelled the world looking at holes in the ground. And the best part about mining tourism, as opposed to common or garden tourism, is that I got paid to do it. Thankfully, I have never developed the weak-at-the-knees kind of love of the pit that the Geos seem to have.

---

[1] There are a whole bunch of jobs in mining that are geology related (rocks are important in mining):

- o   geologists (look at what rocks are where, generally by drilling lots of holes and analysing what comes out – most geologists have spent years as graduates on drill rigs in a desert logging drilling results – they all talk about it);
- o   geophysicists (use funny methods – not drilling, to find which rocks are where);
- o   hydro-geologists (underground water);
- o   geotech (how steep can you cut a cliff in the rock while making sure that it doesn't collapse);
- o   mine planners (in what order you dig the rocks out); and
- o   metallurgy (how do you process the rocks in order to sell them).

All these jobs still look pretty similar to me and, much to the disgust of the geo-scientists, I tell them that I can't tell the difference between them. Even though the term 'Geo' generally only refers to actual geologists, in this text I shall refer to this whole cohort of geo-scientists as Geos.

I got sent to North America, Africa, Europe and Asia, as well as all the deserts in Australia (and there are a lot of deserts in Australia) to pursue my mining tourism. Eventually I got the 'full house' when I got a gig reviewing a project in South America. I had now 'worked' on every continent except Antarctica (and Antarctica doesn't really count - because of all the ice.) I was no longer exaggerating when describing myself as 'an international man of mystery.'

An important facet of mining tourism is frequent flyer points (which are a ridiculous thing really if you do a lot of flying. If you fly so much that you're sick of it, you can get enough points to fly more). The thought of flying on a plane without gaining a point was devastating.

Once I got to fly to Europe first class! (Being a simple, colonial boy, I'm not cut out for first class travel. Before the flight, the first-class hostess explained that there was an executive chef for just the eight people in the first-class cabin. The first-class chef came by a little later to explain that if I could dream up any other dish using any combination of the ingredients on the menu, he would be ever so happy to make it. This was very different from economy. I still hoped that the first-class chef would have a better idea of how to best combine the ingredients. That is what I expect of non-airline chefs. He went on to ask at what times during the flight would I like to eat. This was getting silly now, surely, he was the expert at optimal flight eating times. As it turned out, I woke up earlier than the others on this flight. It was dark and all the other first-class cabin doors were shut. But I was awake and I was getting hungry. I pressed my bell and the first-class hostess magically appeared a minute later. I asked if it would be ok to have breakfast early. In a first-class voice she said "Certainly how would you like your eggs cooked?"

"Poached."

"I'll wake the chef!"

"Oh, it's not that important, any other way is fine, scrambled or fried or boiled, I'm not that fussy."

"No, I'll wake the chef." And she was gone. The chef arrived soon after and my poached eggs soon after that.)

If you get used to first class, then you're too rich for your own good.

Travelling to West Africa from Australia is a bloody long way. We flew to Dubai and then got on a plane that did a twice weekly round trip from Dubai to Conakry (in Guinea), then onto Dakar (in Senegal) before returning to Dubai. Its only when you're flying across the top of Africa that it becomes apparent just how big it is. Long flights give plenty of hours to read project documentation. (You normally get about 1000 pages to read before a review. It's important to flip through and look at all the pictures and graphs. This is normally sufficient to have enough background to ask some salient questions when the review starts.) Arriving in Guinea there was a forty-minute period where we transferred between the business class cabin on the plane to the compound at the Sheraton Hotel. During this time, I was undeniably in the third world. My middle class, university educated upbringing did nothing to prepare me for it.

The next day we travelled to the mine site by a much smaller plane. It met the company aviation standard: two engines, two pilots and two wings. Pilot number one was an older Belgian gentleman. Pilot number two was from Alabama and spoke with an undeniable Alabama accent. He did the safety briefing which was very brief and included instructions like: "If I jump out of the plane, then you'll probably want to follow me." I was a bit apprehensive at first. But my second thought was that this guy seemed like a survivor. I picked a seat behind him.

When we arrived at the mine site, we began our meetings in the conference room over lunch. By the time the presentation started, we were jetlagged and well fed. The lights were dimmed and I'm sure the heater was turned up. My eyes were very heavy and then the session on Community Engagement started and just like that I was out. To make things worse I was sitting at the end of the table closest to the screen. So, everyone would have seen me sleeping. I hope that I didn't snore too loudly to highlight my predicament. They were all polite enough to not mention it, or maybe everyone else was also napping too.

When we got back to our mine site accommodation (which was three of us in a three-bedroom house), I did what I always do in a hotel room. I made a small pile of my underwear and socks on the floor, before going to bed. My intention was to add a small amount to the pile the next evening and then collect it all into my washing bag the following morning before flying out. (Never let yourself get separated from your bag on fly out day.) When I got back to the room that evening, to my surprise (along with an uneasy meld of disgust and delight) I found my jocks and socks washed, ironed and folded on my bed. My first thought was that I shouldn't have left them on the floor because some poor cleaning lady had then felt obliged to wash them for me. If I was less of a slob I could have put them in my little washing bag last night and not inconvenienced her. My second thought was that if I had known how things go down in Africa, I could have emptied my whole washing bag on the floor last night and it would all be clean now! Since we were flying out the next morning, it was too late to take further advantage of the cleaning regime.

When going through Frankfurt airport with a review team on our way home, my carry-on bag set off the machine that goes 'bing.' I don't know why because I don't speak German. They re-scanned my bag and sure enough it was still 'binging.' I was kept aside for a couple of minutes until a huge German policeman appeared on the scene. He was the size of Arnold Schwarzenegger. He stood a full head taller than me and must have weighed a fit 120kg. He carried a baton in one hand. He spoke to me in German. I pulled a funny face and shrugged as I didn't understand. He made me stand still with my arms out as he waved a scanner around my body. I was pretty relaxed; I didn't have any dangerous goods so surely this will all blow over soon and I'll be on my way. Maybe I was too relaxed. While my arms were still outstretched, he grabbed me by the belt buckle and pulled me towards him almost lifting me off my feet. This was a real power move by the big German cop. If I was too relaxed before, I was figuratively crapping my dacks now. I stood there for another few moments hoping to be released. Then he waved me off. I couldn't grab my bag and get out of there fast enough. While I was being accosted, the rest of the review team were pointing and laughing (at least they didn't just leave me, I guess). One of the guys thought it was so good that he took some photos of the ordeal on his phone. As I approached the group, the robo-cop came over and made him delete the photos (they don't like you taking photos in airports). Not wanting to be dragged around by the belt buckle, he complied immediately.

◊◊◊

When I am asked what I do, I say that I get lied to for a living. I sift through the lies to get to the truth. Of course, I am very rarely told actual full-blown lies, but everyone has a perspective (some are helpful and some not so much).

I was told once that management needs consultants. Even when a CEO who knows what he wants to do, sometimes he needs the support (and report) of an external consultant to convince his board (and someone to blame if it turns out bad). I was happy to help.

The strangest job I had was when I was engaged to read resumés. An engineering company that I knew was on a recruitment drive for a new project, which culminated in a stack of about 300 resumés that someone needed to go through to decide who to interview. All of the managers at the engineering company were far too busy (either doing real work or being important) to spend hours resumé reading. I did it at my daughter's grade 1 school disco. I was "that parent" sitting in the corner with my laptop doing billable hours while the kids were dancing.

It's not all beer and skittles – the jetlag can be pretty bad. Sitting through presentations on the first day after a long-haul flight is tough. Typically, it's just after lunch, in the warm, dark presentation room, that they start talking about geotech. This is the danger zone. Geotech is colossally boring. There are reams of graphs that make no sense and the persistent tug of your home time-zone. The best thing to do is to stand at the back of the room (pretend that you have a bad back if you need to justify standing). Drink a coffee (which is difficult in the USA because their coffee is crap). Rock from one leg to another if you need to. Just do not stay in your seat!

At moments like these, there is a direct correlation between remaining in your seat and sleeping. I've been in this situation so many times with my mind screaming "GET UP! GET UP!" while my body replies in a soothing velvet voice "but it's so comfortable here, it's warm and dark. I don't know anything about geotech, and that's not about to change even .. if ... I .... listened ....." I have tried to rest my face on my hands with my elbows on the table in a way that looks like I'm deep in thought. There are two risks with this approach:

1. If I were asked a question, I would only be able to snore in response; and
2. If my cunning sleep plan were to be fully successful, my head would inevitably roll off the hand-stands and on to the table.

Either way, the game would be up. It is very difficult to take decisive action when all you want is closed eyes. But you must. Do not remain in your chair. No-one wants a project reviewer asleep at the end of the table. Trust me, just like cutting off a finger in a lawnmower, it's a mistake that smart people only make once. Really smart people manage to avoid it altogether.

I have since discovered melatonin. Even a small dose of 3g of melatonin has meant that I can sleep during the night when arriving in a new country and pretty much avoid bad jetlag. Consequently, I haven't fallen asleep in a meeting for almost a decade.

One day I was at home writing reports. I did my usual procrastinating. I made coffee. I took the kids to school. I walked the dog. I mowed the lawn. Eventually, I approached the dining room with my laptop to find Bella already at the dining room table typing on her computer (Bella was also working from home that day). I stood there confused and unable to clearly express my emotions. She looked up, and I stammered, "Um, Babe ... you're in my seat."

"There are six chairs around the table. Can't you use any of the other ones?" She made no sense at all.

"That's the seat that I sit in when I'm writing my reports." There was a standoff for a minute until I decided to stop being a baby and sat down on another chair. I made a mental note to make sure that I was first to the table whenever Bella was working from home.

The ultimate aim while consulting is to have a day that you can charge for all 24 hours (or more). To achieve this, you really need a special set of circumstances. It would have to include something like doing a job (on day rates – nominal ten hours) for Client A during the day. Then in the evening, catching a long-haul flight to travel to Client B (on day rates – nominal ten hours). And while on the plane do some work for Client C (need to do four hours of report writing or reading). I have never managed to achieve this state of consulting-nirvana, but you can see that I have thought too much about it already. Rest assured that if three jobs for three different clients coincide in just the right way, I'll be ready!

How long will it last, the life of an international mining tourist? I don't know. But if the pattern of my career is anything to go by, the next thing will be better than the last one. If the phone keeps ringing, I'll keep saying yes. If the phone stops ringing, I guess I'll start applying for real jobs.

Sometimes I get asked how do I manage the uncertainty. Generally, I know what bookings I have for about two months ahead. Beyond that there is a whole lot of empty space (or opportunity). The price of non-exclusivity seems to be uncertainty. And I love it.

EPILOGUE

It was getting late. The outdoor fire had almost burned out. The moon was high in the starry sky. The other guests had started departing.

"...and that, Wranger, is what I do."

"Well, that was a fair bit more information than I was expecting." Wranger replied. "If you just said that you're an Engineer, we could both have assumed that your job was boring and not talk about it. But there's one thing that I can't figure out, Jack."

"What's that, Wranger?"

"Are you clever or just lucky?" he asked.

"I'd rather be lucky than clever."